Decolonizing Development

Antipode Book Series

General Editor: Noel Castree, Professor of Geography, University of Manchester, UK
Like its parent journal, the Antipode Book Series reflects distinctive new developments in
radical geography. It publishes books in a variety of formats – from reference books to works
of broad explication to titles that develop and extend the scholarly research base – but the
commitment is always the same: to contribute to the praxis of a new and more just society.

Published

Decolonizing Development: Colonial Power and the Maya
Joel Wainwright

Privatization: Property and the Remaking of Nature–Society Relations
Edited by Becky Mansfield

Cities of Whiteness
Wendy S. Shaw

Neoliberalization: States, Networks, Peoples
Edited by Kim England and Kevin Ward

The Dirty Work of Neoliberalism: Cleaners in the Global Economy
Edited by Luis L. M. Aguiar and Andrew Herod

David Harvey: A Critical Reader
Edited by Noel Castree and Derek Gregory

Working the Spaces of Neoliberalism: Activism, Professionalization and Incorporation
Edited by Nina Laurie and Liz Bondi

Threads of Labour: Garment Industry Supply Chains from the Workers' Perspective
Edited by Angela Hale and Jane Wills

Life's Work: Geographies of Social Reproduction
Edited by Katharyne Mitchell, Sallie A. Marston and Cindi Katz

Redundant Masculinities? Employment Change and White Working Class Youth
Linda McDowell

Spaces of Neoliberalism
Edited by Neil Brenner and Nik Theodore

Space, Place and the New Labour Internationalism
Edited by Peter Waterman and Jane Wills

Forthcoming

Grounding Globalization: Labour in the Age of Insecurity
Edward Webster, Rob Lambert and Andries Bezuidenhout

Decolonizing Development

Colonial Power and the Maya

Joel Wainwright

Blackwell
Publishing

BLACKWELL PUBLISHING
350 Main Street, Malden, MA 02148–5020, USA
9600 Garsington Road, Oxford OX4 2DQ, UK
550 Swanston Street, Carlton, Victoria 3053, Australia

First published 2008 by Blackwell Publishing Ltd

1   2008

Library of Congress Cataloging-in-Publication Data

Wainwright, Joel.
  Decolonizing development : colonial power and the Maya / Joel Wainwright.
    p. cm. – (Antipode book series)
  Includes bibliographical references and index.
  ISBN 978-1-4051-5705-6 (hardcover : alk. paper) – ISBN 978-1-4051-5706-3 (pbk. : alk. paper)   1.   Mayas–Belize–Toledo District–Economic conditions.   2.   Mayas–Belize–Toledo District–Economic conditions.   3.   Mayas–Agriculture–Belize–Toledo District.   4.   Mayas–Belize–Toledo District–Social conditions.   5.   Toledo District (Belize)–Colonial influence.   6.   Toledo District (Belize)–Economic conditions.   7.   Toledo District (Belize)–Social conditions.   I.   Title.
  F1445.W35 2008
  305.897′427072824–dc22

                                                    2007026743

A catalogue record for this title is available from the British Library.

Set in 10.5/12.5pt Sabon
by SPi Publisher Services, Pondicherry, India

For further information on
Blackwell Publishing, visit our website at
www.blackwellpublishing.com

For Julian Cho –
nim li winq, in wami:
Bantiox aawe

Contents

List of Figures

Acknowledgments

The research for this book was made possible by the financial support of a Fulbright fellowship, a grant from the MacArthur Interdisciplinary Program on Global Change, and fellowships from the University of Minnesota. A Killam postdoctoral research fellowship provided time to write at the University of British Columbia; the book was completed at the Ohio State University. I would like to thank the people of Minnesota, British Columbia, and Ohio for building these great public universities where I have had the good fortune to study, teach, and write.

In a substantive way the research for this book started in 1995, and I owe a tremendous, unpayable debt to the many friends and allies who have sustained and inspired me in the subsequent twelve years. To the following I offer a sincere, and simple, thank you: Daniella Aburto, Pedro Ack and his family, Basilio Ah, Mazher Al-Zoby, Kiran Asher, Nurcan Atalan-Helicke, Luis Avila, Trevor Barnes, Bruce Braun, Jon Brier, Matalia Bul, Oscar Burke, Domingo Cal, Padraig Carmody, Noel Castree, Maggie, Ian, and Ingrid Cho Absher, Ines and Pio Coc, Mateo, Maria, Pas, Marta, Mon, and Cristina Coc, Andres Coh, Mat Coleman, Bud Duvall, Jerry Enriquez, Peter Esselman, Vinay Gidwani, Jim Glassman, Glenn Gould, Leila Harris, Qadri Ismail, Allen Issacman, Carmelito Ixim, Will Jones, Brian King, Premesh Lalu, Gomier Longville, Josh Lund, Geoff Mann, Anant Maringati, Jacob Marlin, Mayday Books, Bill and Diane Mercer, Peg and Hans Meyer, Matt Miller, Jane Moeckli, Bernard Nietschmann, Rafa Ortiz, Tom Pepper, Paul Robbins, Morgan Robertson, Eugenio Salam, Abdi Samatar, Anna Secor, Eric

Sheppard, Adam Sitze, John Skelton, Michelle Spencer-Yates, Karen Steigman, Amanda Swarr, Pulcheria and Bartolo Teul, Mary Thomas, Latha Varadarajan, Dave and Camille Wainwright, Lou and Shay Wainwright, Rick Wilk, Theresa Wong, and Charles Wright.

My most heartfelt thanks are reserved for Kristin Mercer.

List of Abbreviations

AI	Atlantic Industries, Ltd.
BDDC	British Development Division in the Caribbean
BEC	Belize Estate and Produce Company
CARD	Community-initiated Agriculture and Resource Development project
CDB	Caribbean Development Bank
CO	Colonial Office
CRFR	Columbia River Forest Reserve
DC	District Commissioner
DFC	Development Finance Corporation
FD	Forest Department
FPMP	Forest Planning and Management Project
GOB	Government of Belize
IACHR	Inter-American Commission on Human Rights
IBRD	International Bank for Reconstruction and Development
IFAD	International Fund for Agricultural Development
IFI	Intermediary Financial Institution
ILRC	Indian Law Resource Center
IMF	International Monetary Fund
JCS	Julian Cho Society
MAFC	Ministry of Agriculture, Fisheries, and Cooperatives
MFS	Maya farm system
MLA	Maya Leaders Alliance
MP	Minute Paper
NGO	Non-Governmental Organization
ODA	British Overseas Development Administration

OED	Oxford English Dictionary
PUP	People's United Party
RLC	Reservation Lands Committee, or the Committee
TAA	Toledo Alcaldes Association
TCGA	Toledo Cacao Growers Association
TMCC	Toledo Maya Cultural Council
TRDP	Toledo Research and Development Project
TSFDP	Toledo Small Farmers' Development Project
UDP	United Democratic Party
UN	United Nations
WTO	World Trade Organization

Introduction

Capitalism qua Development

Traditional geography steals space just as the imperial economy steals wealth, official history steals memory, and formal culture steals the word.

Eduardo Galeano (2000: 315)

In its brief history, global capitalism has created a world of such intense inequalities that one can only conclude, to borrow Galeano's words, that the world is governed by an imperial economy designed to steal wealth from the poor. Consider: in 2001 the gross net income (GNI) for the entire world was 31.4 trillion US dollars.[1] If this vast sum was distributed equally among the world's 6.1 billion people, it would amount to $5,120 per person. But the vast majority of people in the world received considerably less. In Latin America and the Caribbean, for instance, the GNI per capita was $3,280; in South Asia, $460; in Sub-Saharan Africa, only $450. Such regional averages are deceptive, however, because each of these regions is in turn divided by inequalities that parallel the global pattern, and the subaltern majorities do not own (let alone earn) even these modest sums. Thus, in a world with a per capita GNI of more than $5,000, there are 2.8 billion people – almost half of the world – who live on less than $700 a year. Of these, 1.2 billion people earn less than $1 a day. This is much worse than it was a generation ago. The average GNI of the richest 20 countries today is 37 times that of the poorest 20, a degree of inequality that has roughly doubled in the past 40 years.[2]

The irony is that this historic expansion of inequality occurred during a period known as the "age of development," a time when "development decades" came and went and scores of states built their hegemony, along with multilateral institutions and NGOs for that matter, upon one mandate: *accelerating development*. A truly global consensus emerged concerning political-economic management – a form of hegemony in Gramsci's sense[3] – that the world's poor should enjoy the fruits of development. The fact that global capitalism has increased inequality without substantially reducing poverty raises stark questions: what is it that makes some areas of the world rich and others poor? How is it that capitalism reproduces inequality in the name of *development*? Indeed, how is it that the deepening of capitalist social relations comes to be taken *as* development?

Contesting Development

This book clears space to answer these questions by investigating colonialism and development through the lens of a postcolonial Marxism and by considering the colonization and development of the region known today as southern Belize. This area, also called the Toledo District, is the poorest in the country and among the poorest regions in Central America. The 2002 GNI for Belize was $2,960.[4] The greatest poverty is concentrated in the rural Maya communities in the Toledo District, where 41 percent of the households earned less than $720 per year.[5] For the World Bank as much as the local farmers who experience the existential effects of this poverty, the solution to this situation is economic development via neoliberal policy and loans of financial capital.[6]

The 1990s were a tumultuous decade in the Toledo District of southern Belize as export-oriented neoliberalism became Belize's de facto development strategy. State spending had been governed by a strict austerity and the state privatized public assets at a rate that left it with little left to sell.[7] This complemented a vigorous search for new exports, which have led to an expansion of resource extraction, particularly in fisheries, timber, and agriculture. When the Ministry of Natural Resources sold a number of new logging concessions in Toledo in the mid-1990s, the neoliberal development model collided with an indigenous movement that was gaining ground throughout southern Belize.[8] This social movement – called simply "the Maya movement" in Belize – was led by the late Julian Cho, a schoolteacher who was elected to the chairmanship of the movements' central organization, the Toledo Maya Cultural Council (TMCC), in 1995. Julian and the TMCC struggled to organize Mopan and Q'eqchi' Maya-speaking people, whose livelihoods are based on corn and rice production in the forests of Toledo, to win secure rights to the lands that were threatened by the logging concessions.[9] This Maya movement used the logging concessions as a way to articulate claims about land rights and the marginality of the Mayas in Belizean development on national and international scales.[10]

The drive to expand logging exports and the rise of the Maya movement collided in September 1995 when a logging concession was granted to a multinational firm to cut timber in the Columbia River Forest, an area used by a number of Maya communities for hunting, farming, and collecting other non-timber forest products. Demonstrations by Mayas and their allies called for an end to foreign logging operations, secure land rights, and a new investment by the state in a development project in the region (called "CARD": see chapter 2). To map their territory and present an alternative vision of development, the leaders of the Maya movement organized a project to map all of the Maya communities in southern Belize (I discuss this project in chapter 6).[11] The maps and the logging concessions were two key pieces of evidence in a lawsuit drawn up against the state and brought before the Supreme Court of Belize in 1997. The Maya movement won some of its demands. Logging operations were cancelled in the Columbia River Forest in mid-1996. Maya leaders were invited to assist in designing a new development project, funded by the state with loans from regional development banks, that aimed at improving incomes in rural communities. After the 1998 election of the progressive People's United Party (or PUP) government of Said Musa, "friendly settlement talks" were established between Maya leaders and state representatives to resolve the land issue.

But the Maya did not win all that they had struggled for. Julian Cho died under mysterious circumstances in December 1998. As the movement fractured, the Musa government found that there was no unified leadership and no substantive proposals to negotiate. The settlement talks on the land issue soon dissolved. Today, the same logging company is at work in Toledo's forests; CARD, the development program, has come and gone, leaving Belize with more debt, and poverty has only deepened in the Maya communities. As for the lawsuit, in 2003 the Inter-American Commission on Human Rights (the IACHR, part of the inter-American system of international law) ruled in favor of the Maya, but as of September 2007 the practical effects of this ruling have been nil.[12]

This story resounds with those from many parts of the world today. It is a cliché to say that development projects often hurt the poor, women, or other subaltern social groups. The literature cataloging the hybrid ways that neoliberal capitalism has seized and reformed the political sphere (only to be met by new forms of resistance) is vast. As in southern Belize, a common narrative involves environmental threats and conflicts between different social groups and the state that are resolved through a shift from political and legal to developmental policies.[13] Today, threats to hegemony that emerge through such conflict are always already negotiated and resolved in terms of *national development*, a political surface that expands and contracts as hegemony is reworked in struggles over capital accumulation, identity, territorialization, and social power. Though this book examines the politics of development in contemporary Belize, my aim is not simply to document neoliberalism's effects – nor to write an ethnography of the Maya or their resistance.[14] Rather, this is a study of the history and politics of development as a form of power, one with a truly global sway. In the wake of formal, political decolonization, development became the central mission or justification for Third World states. These states faced the enormous challenge of reconfiguring long-standing economic patterns and processes that were immiserating much of the world.[15] The promise of development has gone unfulfilled for most of the world, and we must criticize the development policies that have failed to create the conditions for local capital accumulation, social investment, or sustainable livelihoods.

This task has been made more urgent in the past twenty years. The disastrous consequences of neoliberalism and structural adjustment, consolidated as the de facto development project for the world, led many to suggest a relationship between imperialism and development.[16] The authority of the Bretton Woods institutions – the IMF, the World Bank, and the GATT/WTO – is vast and plainly rooted in colonialism. For Belize, the transition from colonial rule to neoliberalism was seamless: the government gained formal independence from Britain only in 1981, and in the face of a growing balance-of-payments crisis adopted its first agreement with the IMF in 1985.[17]

Just as there can be no doubt that neoliberalism holds sway in discourses about development and economic management today, there is a parallel strength to the enframing of development issues as the property of nation-states.[18] For instance, the balance of accounts and trade deficit are understood as *Belizean* problems, notwithstanding the facts that the economic life of Belizeans exceeds the territorial extent of the state, and that Belize's elites are increasingly transnational. That the constellation of issues that are thematized as "economic" is defined vis-à-vis the territory of the nation-state is neither innocent nor particularly old. The very identification of "the economy" as having an essentially national character dates from the early twentieth century.[19] At both the local and global scales, the economy has been constituted as a sphere of economic flows regulated by national policies. This formulation of the economic as a geographical object is rooted in the colonial period.

Although this book concerns development in Belize, I do not treat Belize as an unproblematic site of analysis. If we begin by simply assuming that Belize is *there*, if we presume that the ontology of "Belize" is fixed in advance, we stand to miss a crucial effect of colonial power. The iterative production of Belize as a territorial nation-state works through practices that are thoroughly colonial. This is one of the lessons of the Maya land rights movement – what we call "Belize" today is an object produced through Spanish and British colonialism. This process of becoming Belize cannot be disassociated from primitive accumulation and the production of essentialist forms of national and racial forms of subjectivity. These effects are reiterated in the colonial present through the very act of taking Belize as an unproblematic object.[20] Like much of the world, the processes that have played the greatest role in shaping the political economy and social life in Belize are both colonial and capitalist; therefore I focus on these relations. To interpret them effectively requires an engagement between development and the Marxist and postcolonial traditions.

Nature/Development

In *Keywords*, Raymond Williams argues that "nature" is "perhaps the most complex word in the [English] language"[21] because it gathers three radically different meanings under one sign. "Nature" can refer, first, to the essential quality of some thing. If we ask after the nature of a thing, we are asking after its essence. Second, "nature" can refer to an "inherent force which directs either the world or human beings or both"; third, "nature" can also refer to the world itself, environment, the space in which things live. These meanings are frequently conflated when some thing is described as being "natural." An affiliation between essence, direction, and environment is thus woven through our language. Williams explains of "nature":

> What can be seen as an uncertainty was also a tension: nature was at once innocent, unprovided, sure, unsure, fruitful, destructive, a pure force and tainted and cursed. The real complexity of natural processes had been

> rendered by a complexity within the singular term.... The emphasis on discoverable laws...led to a common identification of Natural with Reason: the object of observation with the mode of observation.... Each of these conceptions of Nature was essentially static: a set of laws – the constitution of the world, or an inherent, universal, primary but also recurrent force...teaching a singular goodness.[22]

Fruitful yet destructive, a pure force and yet tainted: synonyms of "development," an equally difficult keyword that Williams, alas, did not define for us in *Keywords*. Our inherited concept of "development" shares much in common with "nature." Like nature and culture, development is one of those words that first described "a quality or process, immediately defined by a specific reference, but later became independent nouns." Also like nature, development carries multiple and radically divergent meanings. The first is the *unfolding* of something essential, as in "plant development" or "child development." This is the older meaning – older even than the English word "development." The verb "to develop," from which "development" is derived, has Latin roots that carry the connotation of "disentangling." "Development" thus refers to a particular ontological quality that is expressed through the process of unfolding.[23] Aristotle in *Physics* uses the illustration of the seed to speak of the essence that is expressed in the totality of its unfolding. Here is Aristotle in Book IV of *Physics*, chapter 1:

> We also speak of a thing's nature as being exhibited in the process of growth by which its nature is attained. [This is "development" as ontology, i.e., unfolding of (the) latent.] ... But it is not in this way that nature (in the one sense) is related to nature (in the other). What grows *qua* growing grows from something into something. Into what then does it grow? Not into that from which it arose but into that to which it tends. *The shape is then nature.*[24]

Thus the essence of nature as essence is given in what – today – we would call development. That term was not available to Aristotle, or, for that matter, anyone before the 1800s. Not before the rise of the nation-state-capital trinity: a clue to our inquiry. The modern usage enters Western philosophy via Hegel, who defines development with the example of the seed developing into a plant in his *Encyclopedia*. Hegel usually uses "development" in the ontological sense, i.e., to refer to the self-unfolding of life toward the divine or of "the divine in the world."[25]

Second, "development" also refers to an intention to create or change something.[26] In this sense, "development" refers to a force that tutors a change in something or a course of events. This meaning always carries the sense of *will*: development in this second sense implies an intervention – to make something move in a direction that is *not* given in advance, essential, or required. The object of development is changed, moved, or improved, by some willful power applied from above and outside of it.

Our concepts of "development" and "nature" share this problematic conflation for a common reason: they are two of our most entrenched, inherited, ontological signs for indicating essence. In Western metaphysics "nature" and "development" both express essence by proposing a relationship between temporality, spatiality, and ontology. As with nature, development is sometimes defined as an inherent force which directs human beings. Nature binds temporality and ontology by joining worldliness as totality with interior, substantial essence. The substantiality of nature articulates interiority and becoming: for instance, again, in Aristotle's *Physics*, Book II, we read: "nature is a source or cause of being moved and of being at rest in that to which it belongs primarily, in virtue of itself and not in virtue of a concomitant attribute."[27] Nature is perhaps an older concept than development, but we can see its relation to development in Aristotle's claim that nature is a "cause of being moved . . . in virtue of itself." The essence of nature is expressed through development. Development thus binds temporality and ontology via the *rational unfolding of presence*.

The distinct meanings of development are frequently conflated in ways that have important effects. When we refer to "national economic development," for instance, we at once refer to something that is desirable, that requires willful intervention, and also is a "natural" thing for the nation to do. This conflation is not due to a choice made by the speaker. It is an effect of language – and one of great significance. To consider the implications of this, we need only add two additional comments. First: it was precisely the promise of "national economic development" that every state promised its people on the eve of independence, and it is the global and structural failure to deliver on this promise that animates all our discussions of development today. Yet though we may recognize the globality of this failure, everywhere it remains the remit of the *nation-state* to resolve. Second comment: today, "national economic development" always refers to the deepening of *capitalist* social relations, even when it is not named as such. This affiliation between capitalism and the compound sign "development" has fundamental political effects. The unfolding of capitalism on an ever-wider scale – a process driven by the contradictions of capitalism as a mode of production – is inscribed with an undeserved sense of *directionality.* This directionality may be historical (in the sense of "inevitability"), spatial (in the sense that it produces spatial relations that are taken for granted), or ethical (by implying guidance towards ends desired by liberal-humanist values).[28] Very often these are combined in ways that make the worldliness of the world seem like a "natural development."[29] When capitalism is treated *as* development, the violent effects of the capitalist social relations are normalized and unjust geographies become hegemonic. The "historical identity between Reason and capital" assumes its epistemic and ontological privileges when the extension of capitalist social relations is taken *as* development. Thus of development we could say what Adorno once wrote of "progress": "one cannot employ the concept roughly enough."[30]

The Post-Development Challenge

The failing of the best-known Marxist approach to destroy development conceptually – I am speaking of the political economy of development tradition[31] – led to the rise of the "post-development" school. This group argues that development cannot be understood outside of, or prior to, its operation through discursive practices.[32] To its credit, this move reopens *the* fundamental question of development studies: *what is development?*

Within this literature, the general answer that has been provided is that "development" is a discursive formation exported via global institutions in the mid-twentieth century, extending from centers of power through the Global South via development projects. In a widely read case study, James Ferguson argues:

> "Development" institutions generate their own form of discourse, and this discourse simultaneously constructs Lesotho as a particular kind of object of knowledge, and creates a structure of knowledge around that object. Interventions are then organized on the basis of this structure of knowledge, which, while "failing" in their own terms, nonetheless have regular effects ... [including] the entrenchment of bureaucratic state power, side by side with the projection of a representation of economic and social life which denies "politics" and ... suspends its effects.[33]

Because development's gravitational pull on politics encourages centralized forms of leadership and favors the "developed" over the "underdeveloped," uneven power relations and the authority of the bureaucratic state are deepened in the name of development.[34]

Numerous criticisms have been leveled against the post-development literature.[35] Two are especially pertinent. First, "development" has often been reduced to a singular, monolithic discourse, devoid of any contingency.[36] Ironically, in their effort to displace "development," the post-development critics have often implied that development is essentially singular, and that it has been so since its inception ("in the early post-World War II period" according to Arturo Escobar).[37] That is, for a project that aims at showing, again in Escobar's words, "how the "Third World has been produced by the discourses and practices of development,"[38] the work treats development as monolithic. Yet as Vinay Gidwani writes:

> To proceed, as post-development scholars do, on the assumption that "development" is a self-evident process, everywhere the same and always tainted by its progressivist European provenance . . . is to succumb to the same kind of epistemological universalism that post-development theorists . . . are at such pains to reject.[39]

Second, critics have shown that there is a notable weakness within the literature that I would call, following Gramsci, the "analysis of situations": careful studies of class formations, production and consumption, and state-society relations. On this point, Michael Watts argues that post-development is weakest where it matters most. Escobar and colleagues fail to adequately analyze how development discourse is articulated through concrete socioeconomic practices; Escobar's work, Watts once remarked, is insufficiently dialectical.[40] To capture the subtleties of that dialectic, Watts called not for post-development but rather "development ethnographies."[41] Yet our challenge is not ethnographic. Certainly, discerning the effects of development practices presupposes a rich understanding of state-society relations, and we must examine the sedimented effects of the historical-geographical processes that have shaped the particularities of capitalism qua development and its hegemony. But that is where the similarities with ethnography should end. If we wish to carry out that work under the sign, "ethnography" – surely one of the signature colonial disciplines – we will only introduce more confusion and epistemic violence.[42]

We should therefore leave the term "post-development" behind. The "post-" before development serves only to draw us off the path of the inquiry. Unlike postcolonialism, which is a concept that I will take up and argue for, in the end, "post-development" amounts to little more than the facile negation of the object it criticizes.[43] Instead of "post-development," we need a fundamental critique of development: one that examines its power, its sway, as an aporetical totality. What is needed, I argue, is a specifically *postcolonial* Marxist critique of development. It is notable that post-development failed to incorporate Marx's critique of capitalism and failed to incorporate postcolonialism. Yet a postcolonial Marxism that rethought development would retain two key points from the post-development literature. First: a critique of capitalism must have the theoretical tools to take apart "development" on discursive and ontological grounds. This clarifies how we can leverage what counts as "development" away from its historical moorings as *trusteeship*.[44] Second: the reading that produces this critique cannot assume an *a priori* alternative (so-called "real development" or "true development") that stands apart from the critique. The latter distinguishes post-development from earlier anthropological critiques of development that rested on the argument that capitalism's failure was to displace holistic cultural systems that were the only real basis for "true development."

Development as Aporia

In the wake of formal, political decolonization, development became the central mission of and justification for Third World states. These states faced the enormous challenge of reconfiguring longstanding economic patterns and processes that were immiserating much of the world. The promise of development has gone unfulfilled for most of the world, and we should criticize the policies that have failed to create the conditions for local capital accumulation, social investment, and sustainable livelihoods, not to mention healthy, long lives – in the name of *development*.

This reading, inspired by an unholy alliance of post-Enlightenment critics (Marx, Derrida, Spivak, and others) aims at doing something other than *rejecting* development. We cannot *not* desire development.[45] Development remains an absolutely necessary concept *and also* absolutely inadequate to its task. In this way – insofar as development is a necessary but impossible concept for understanding capitalism – development constitutes an *aporia* for postcolonial Marxism. Spivak distinguishes aporias from dilemmas and paradoxes by the way they are "known in the experience of being passed through, although they are non-passages; they are thus disclosed in effacement, [as the] experience of the impossible." An example, which in fact closely parallels the aporia of development, is liberal law: "'Law is not justice, [although] it is just that there be law,' says [Derrida in his essay] 'Force of Law' ... Justice cannot pass in a direct line to law; that line is a non-passage, an aporia."[46] I add that aporias, in their impossible passage, may be productive in ways that paradox and dilemma are not.

To be precise, the aporia here consists in the radical doubt encountered by all those who would wish to criticize development. On one hand, "development" is a site of great epistemic violence; on the other, development remains absolutely necessary for us – since it is, in Spivak's words, "the dominant global denomination of responsibility." Spivak wrote these lines in the late 1990s, when neoliberalism was the global development strategy. As the denomination of responsibility, development is *undignified*. She continues: "The story is that the rich nations collectively hear the call of the ethical and collect to help the poor nations by giving skill and money."[47] It is a "story," indeed, produced and told at the cost of considerable labor. Yet insofar as we should in fact solicit the call of the other, we need to somehow transform the "dominant global denomination of responsibility." The problem is that the denomination, the currency, is no good. We cannot pay the debts of our responsibility with development dollars. In *The gift of death*, Derrida writes:

> Such is the aporia of responsibility: one always risks not managing to accede to the concept of responsibility in the process of *framing* it. For responsibility...demands on the one hand an accounting, a general answering-for-oneself with respect to the general...and, on the other hand, uniqueness, absolute singularity, hence nonsubstitution, nonrepetition, silence, and secrecy.[48]

But why do we have responsibility at all? Quite simply because of the abject facts of poverty and inequality. We live in a world where billions of people do not have sufficient food and clean water. *Rejecting* "development" – the hegemonic denomination for our responsibility – is neither morally possible nor desirable. Thus there can be no simple negation or rejection of development. Not because development is good (it is not), but because a rejection still turns within the analytic space opened and shaped by development discourses. Development marks the site of a fundamental doubt that must be struggled through in order to produce stronger positions and concepts. Our challenge is to do so without being seduced into its sway, where in the face of injustice, inequality, and expropriation of value we simply ask how to "improve" or "accelerate" development. Insofar as capitalism qua development cannot change the basic conditions of inequality and the extraction of surplus value from labor, its embrace should be rendered impossible, that is, impassable.

Within this view, what must be examined and explained is the articulation of capitalism with development in its dual sense. This articulation is neither transhistorical, nor aspatial, nor apolitical; it emerges during the period of industrialization in Northern Europe, and – contrary to an argument made by Arturo Escobar, that "development" was "constructed" after World War II – I argue, in concert with Timothy Mitchell, that "development" (as it emerged in the early nineteenth century) was extended through European colonial practices, which, I suggest, called for capitalism to take up an ontological attachment with development. The theoretical difficulty for us is that development consequently became, via imperialism, not *dialectically* but *aporetically*, both inside and outside of capitalism. Development, again, is capitalism; yet it also exceeds capitalism and names a surplus necessary to the correction of mere capitalism. For Cowen and Shenton, this is what constitutes doctrines of development, the need to solder these two conceptions together in historically specific ways. But their logic does not go far enough. To borrow an expression from Jacques Derrida, development is a *supplement* to capitalism – it is a historical-geographical process taken to be *outside* of capitalism, and yet something always already *included*, to make it whole, to allow capital to assume a sense of historical purpose and directionality. Consequently it was only development – not civilization, not modernity, not progress – that was universally taken up after the end of colonialism to define and organize the nation-state-capital triad everywhere. Only development enjoys this degree of epistemic-ontological privilege.

I argue that this aporia – development as denomination of responsibility – has its roots in the very formation of capitalism on a planetary scale through the imperial experience. This clarifies our challenge: to think through this aporia without being seduced by it. But how do we reserve our critique of capitalism without falling into the trap of "post-development," that is, of supposing that we can escape this totalizing structure? How can we at once highlight and undermine the conceptual work that development does for capital? How do we reserve analytic clarity that we are speaking of capitalism when we invoke "development," and open development to a more fundamental critique, yet without abandoning the question of development *in toto*?

Capitalism qua Development

My answer to this challenge is to propose the concept of capitalism qua development. The sublime absorption of capitalism into the concept of development has created the effect that capitalism *is* development.

This condition is, I repeat, neither transhistorical nor aspatial; its articulation occurred at a specific time and place; its roots fed by Enlightenment philosophy; its consolidation as one of the fundamental discourses for speaking of and producing the world is a product of European colonialism. Capitalism qua development has proven to be fundamental to the very ordering of the world. The thematization of the world around categories of "development" – spaces divided by nation-states that are "developed," "developing," and "least developed" – is grounded by its settlement as a means of describing the work of capital, i.e., "capitalist development," the concept that capitalism qua development would replace.

Although development is a form of power that works at the scale of the global, it works through particular institutions and practices in ways that are differentiated. Indeed, capitalism qua development could only achieve its world hegemony because the practices through which it has been constituted are historically and spatially contingent. Colonialism solicited development as a way of organizing a form of hegemony appropriate to the expansion of capitalism beyond Europe as well as struggles over the process of territorialization. To put it baldly: development emerged as a global alibi for the imperial extension of specifically Western modes of economy, spatiality, and being. This event occurred when European colonial practices called for capitalism to take up its ontological attachment with development – essentially soliciting capitalism to become development. While this process seems to have started in the mid-nineteenth century during the age of empire, the *possibility* of this attachment was already in Europe's theoretical repertoire, since its roots lie in Enlightenment philosophy and the desire to see an abstract and unsituated reason applied to direct the movement of History towards the good of humanity. "Development" and "reason" are two watchwords of that philosophical event. Criticizing capitalism qua development requires that we call into question these underlying categories – in order to better understand the work of development in reproducing an imperial, and hegemonic, form of power that governs the world.[49]

Insofar as capitalism qua development cannot change the basic conditions of capitalism – the extraction of surplus value by the hegemonic social class, expanding inequality and concentrations of power – its embrace must be *rendered impossible*. Adorno once wrote of progress, "it occurs where it ends." So too development.

Postcolonialism

As a growing literature has demonstrated, the postcolonial problematic is unavoidable for those who wish to understand development.[50] Yet students of development who turn to the postcolonial literature for answers are often disappointed, because postcolonial theory offers no particular development theory or strategy. Indeed, the literature on postcolonial theory is roiled by debates about the term "postcolonial" itself, as well as the nature of its problematic.[51] Though it is beyond the scope of this introduction to survey these debates, I stress that it is postcolonialism, more than poststructuralism or any other post-, that allows us to extend the Marxist critique of capitalism qua development. I therefore underscore my agreement with Qadri Ismail that "the current epistemological or disciplinary moment" is better characterized as post-colonial than poststructuralist, and partly for this reason, the theoretical arguments wrought by postcolonialism must be "(de)fended: fostered, nurtured ... *consolidated*."[52] In this light, this book may be read as an attempt to consolidate, affirm, and abide by postcolonialism's teachings in order to critique capitalism qua development.

In perhaps over-formulaic and didactic terms, let me briefly sketch four of the lessons of the postcolonial literature that have guided this research, and to indicate how these shape the book. The first lesson is that the achievement and perpetuation of colonial rule required the production of particular forms of knowledge. More: colonialism required the production of forms of knowledge which in turn constituted forms of subjectivity and worldliness that facilitated colonial rule. Colonial forms of knowledge, for instance, may produce the other qua ethnos (through representational practices that may well not present themselves as representing practices) that justify racial exclusion and the extraction of value from the colonial periphery.[53] Moreover, post-colonial studies show that colonial knowledges have outlasted formal colonialism and live on in the present, constitute the present as such, and have ongoing political effects (cf. the US invasion of Iraq: among other things, a product of the legacy of ways of representing the "Middle East" as a place of mystery and irrationality, to be subdued).[54]

This argument is rightly associated with the work of Edward Said, who relentlessly pursued the effects of colonialism on cultural practices. In *Orientalism* and related essays Said shows how orientalism works as both a discipline of study about "the Orient" and a name for a mode of colonial knowledge that *produces* Europe's other. Said's argument has its limitations, and it is not obvious what it implies for capitalism qua development – Said, after all, was no Marxist. In Part I, "Colonizing the Maya," I extend the Saidian problematic (a product of the thought of Gramsci and Foucault, who figure prominently in chapters 1 and 2) to consider the ways that Mayanism contributed to the colonization of the Maya. But unlike Said's critique of orientalism, my critique of Mayanism focuses explicitly upon its dynamic relation with the unfolding of capitalist social relations. I think that this is imperative given the imbrications of colonial discourse with imperialism as a capitalist venture. In this way may we discern capitalism qua development as an object for postcolonial Marxism.

A second lesson is that nationalism, as a response to the problems caused by colonial knowledges and discourses, has proven to be insufficient to the task of decolonizing the world.[55] This may seem inconsistent with certain strands of postcolonial thought; after all, the postcolonial tradition draws inspiration from Fanon, who once argued that the first task in the struggle against colonialism is "the liberation of the *national territory.*"[56] I will return to territory in a moment; here, I only note that whereas Marxism could be said to have been founded as a critique of capital as a social relation, postcolonialism emerges from the ruins of national territorialization. As Chatterjee explains: "Nowhere in the world has nationalism qua nationalism challenged the legitimacy of the marriage between Reason and capital. Nationalist thought…does not possess the ideological means to make this challenge."[57] An ideology that speaks in terms of an essential link between territory and nation/people/race (i.e., what the Greeks called ἔθνος, transliterated "ethnos") and territory, nationalism cannot cut the link between reason and capital at the core of imperialism.[58] Chatterjee's analysis draws out the crucial argument here: the contradiction between metropolitan capital and the "people-nation" is not resolved, but only *suspended*, by the hegemony of *national development* – which is nothing except the promise of the capitalist state to justly transform society through an immanent process *taken to be equivalent with historical progress itself*. For every capitalist state reconstituting its hegemony (i.e., seeking to establish a rationality of capitalist state power) in the wake of colonialism, "the marriage between Reason and capital" has taken the form of capitalism qua development; after colonialism, every capitalist state promises development as the balm for the violence of colonialism and the anticolonial struggle. Again in Chatterjee's lapidary terms: "the historical identity between Reason and capital has taken on the form of an epistemic privilege, namely, 'development' as dictated by the advances of modern science and technology."[59] The very globality of this pattern reflects a profound epistemic authority.

One implication of this second lesson is that we should not place great hope in resistance articulated in nationalist terms. Moreover, in this study I draw the implication that we must examine *how* ethnos becomes mobilized by and for capitalism qua development, in order to destructure the epistemic authority of development as "historical identity between Reason and capital." Again, this should clarify why this book is not a "development ethnography," nor another ethnography of the Maya. I rather see it as a Marxist-postcolonial critique of the sustaining power of these concepts (development, ethnos) and their interdigitation as national development.

The third argument that I take from the postcolonial literature concerns subalternity. Postcolonialism teaches us to read with a persistent skepticism towards practices that represent subaltern voices.[60] This skepticism is not so much scientific or empirical as it is political and ethical. The challenge is to become open to subaltern histories and geographies without *speaking for* or contributing otherwise to epistemic violence. Premesh Lalu crystallizes the problem with subaltern-voice retrieval: "to claim that subaltern consciousness, voice, or agency can be retrieved through colonial texts is to ignore the organization and representation of colonized subjects as a subordinate proposition within primary discourses."[61] Therefore, although I analyze colonial power in this book, which has always been resisted by the Maya, I do not attempt to define and analyze Maya resistance as such, or even to use the tools of empiricism to prove the existence of such a thing. Indeed, like Casteñeda in his study of Mayanism and tourism in the Yucatan, this study must "suspend designation of the Maya and resist the seduction to define who or what is Maya, since it is precisely my task to critique and contest how this has been done."[62] Casteñeda argues that the work of the category "Maya culture" is "a function of the complicitous history of discursive practices in which Maya alterity has been appropriated for use in Western construction of what it is to be 'civilized' as measured against non-European-derived social forms."[63] I share this argument and contribute to this project by extending the critique of Mayanism as a form of orientalism.[64] But, again, this is not the same as attempting to correctly represent subaltern Maya resistance. As Gayatri Chakravorty Spivak explains, "'being made to unspeak' is also a species of silencing."[65] The postcolonial critique here is directed at forms of historicism and empiricist social science that aim to retrieve a silenced subject.

Alternatively, Spivak teaches us to analyze the aporias of the colonial present without recourse to essentialism. And this, in short, is the aim of Part II: to investigate in a post-empiricist manner the aporias of capitalism qua development. Following Spivak's readings of Derrida, I argue that the postcolonial condition is one in which capitalism qua development is aporetical. The three chapters of Part II comprise readings of distinct aporia.

The fourth argument that I extend from postcolonialism is more of an extrapolation than a direct borrowing of a clear theme. My argument here is that we must call into question the essential congruence of the trinity that structures the very worldliness of modernity, i.e., the state-nation-territory complex. This argument is a modification and extension of arguments made by several postcolonial scholars who have criticized the orientations ("West and the rest"; "North-South divide"; "Europe and its others") that constitute the modern world as such. Again, colonialism has called for hegemonic commitments to the way we see the world: divided and opposed between Europe/colony and the rest/metropole. The problem is not only that our worldliness is bifurcated, but that this bifurcation reflects one view, and in a way that is represented as a universal one. In this operation, hegemony is doubly geographical: it is constituted on the basis of spatial relations, and such relations become hegemonic as geographies are naturalized and sedimented as common sense through political and cultural practices.

It is somewhat ironic, given the emphasis that Said places on *Orientalism* as a form of imagined geography, that postcolonial scholarship has focused so extensively on the *histories* and not the *geographies* of colonial discourses, nationalism, and the like.[66] Postcolonial scholarship called into question the European subject of "History" and "Anthropology" that ground colonial knowledge. But I would argue that this scholarship still needs to think through the becoming-space of the territorial nation-state. Such ontological analysis is needed to think differently about capitalism qua development. The world that has been presented for knowledge is one divided by spatial units separated into different stages of development; the world is figured as a space neatly separated into nation-states that are moving through a common time at different speeds. In this particular way of framing the world – which requires the Cartesian view of space and time that Heidegger criticizes in *Being and time* – there is no space outside of development.[67] The argument here is not necessarily a Heideggerian one, however. In one of his prison notebooks, Gramsci discusses the problem of the naturalization of spatial relations as part of a discussion on the "so-called reality of the real world":

> In order to understand exactly the possible significance of the problem of the reality of the external world, it may be useful to develop the example of the notions of "East" and "West" which do not stop being "objectively real" even if on analysis they prove to be nothing but conventions, i.e. "historico-cultural constructions." . . . It is evident that East and West are arbitrary, conventional, i.e. historical, constructions, because outside real history any point on the earth is East and West at the same time. We can see this more clearly from the fact that these terms have been crystallized not from the point of view of man in general but from the point of view of the cultured European classes who, through their world hegemony, have made the terms evolved by themselves accepted everywhere.[68]

Note that the claim is emphatically not that the real world does not exist. Gramsci accepts, of course, that East and West stand in a material relation to the diurnal rotation of the Earth vis-à-vis the sun. But even these celestial relations are always already received through signs that only make sense through geographies that reflect the hegemony of "the cultured European classes." The broader point is that there are no geographical concepts, terms, or metrics – including territory, scale, nature, or place – that stand outside of language, history, and politics. And therefore questions about exactly *where* some thing *is*, or what a space *is*, can only be perpetually negotiated, deferred, and contested. In this sense, Gramsci's expression "world hegemony" has a double meaning: in producing colonial hegemony on a global scale, the West reproduces a hegemonic worldliness.

This argument works within a postcolonial purview only if we bracket the suggestion, implicit in Gramsci's argument, that our aim is to build a "point of view of man in general." I take it that there is no space from which the humanist subject can see the world in such a universal or geographically objective position. Rather than writing spaces for a point of view of "man in general," what is called for by postcolonial geography is the permanent critique of the practices that stabilize the singular worldliness of the modern world. The critique of orientalism is therefore a challenge to all fixed orientations, a critique of the argument that the world can be known adequately through empiricism. The problem is not simply the particular geography that unfairly represents the East – though that is of course of immense importance – but rather the very distinction between East and West, between Europe and its others. The roots of orientalism lie in this orientation.

Absent fixed, empirical orientations as reference points, how is it that some thing like "Belize" comes to be recognized as both a place and a territorial nation-state? This question is urgent for the study of development, since the territorial nation-state is the hegemonic scale and frame of analysis. To think through the ontological qualities of the capitalist nation-state and its adherence to capitalism qua development, we need a Marxist-postcolonial conception of territory.

Territorialization

The colonial policy of the capitalist countries has *completed* the seizure of the unoccupied territories on our planet. For the first time the world is completely divided up, so that in the future *only* redivision is possible.
Lenin (1997: 76; italics orginal)

A group of people living on a few acres of land will set up boundaries between their land and its immediate surroundings and the territory beyond, which they call "the land of the barbarians." In other words, this universal practice of designing in one's mind a familiar space which is "ours" and an unfamiliar space beyond "ours" which is "theirs" is a way of making geographical distinctions that *can be* entirely arbitrary. I use the word "arbitrary" here because the imaginative geography of the "our land-barbarian land" variety does not require that the barbarians recognize our distinction. It is enough for "us" to set up these boundaries in our own minds; "they" become "they" accordingly, and both their territory and their mentality are designated as different from "ours."
Said (1979: 54; italics original)

A Marxist and postcolonial theory of territory is possible because of the radically different theoretical contributions of Lenin and Said, for whom imperialism and the division of territory are always interrelated processes, and also because of those Marxist geographers who, reading the texts of Henri Lefebvre, have provided a vocabulary to analyze the production of space at the center of political-economic analysis. In his 1978 text, *De l'Etat IV*, Lefebvre argues that the state "binds itself to space" in three ways. The first of these moments is

the production of a space, *the national territory*, a physical space, mapped, modified, transformed by the networks, circuits, and flows that are established within it. ... Thus this space is a material – natural – space in which the actions of human generations, of classes, and of political forces have left their mark, as producers of durable objects and realities.[69]

I join other political geographers in following Lefebvre's lead to investigate how the national territory is produced. The challenge here is to decolonize the realist framework that simply presupposes the natural existence of territory, without asking after its ontological basis.[70] In this view, territory is "the spatial organization of persons and social groups" within the boundaries of the modern nation-state.[71] But the concept of territory (understood as the coordinate space occupied by a nation-state) is a historical-geographical phenomenon of recent foundations.[72]

Postcolonial theory provides an opening for us to radically rethink territory. Consider the epigram by Said,[73] where he uses territory as an example of an object that is created through colonial discourse. Said asserts that the geographical or spatial "fix" of territory must involve the production of sociospatial difference. Yet Said often seems to suggest that the production of territorial difference is transcultural and transhistorical (see the passage quoted at the introduction to this section, describing the demarcation of territory as a "universal process"), a view that runs counter to his thought. Said also suggests that the production of subjects through nationalism – that is, the making of the "us" and "them" of national identity – stands in a close relation to the production of territorial geographies (albeit one that is ultimately arbitrary). While this constitutes a postcolonial entry to a discussion of territory, we must acknowledge that they do not yet explain how territory is materialized, or how exactly these crucial objects (national territories) get made "in our own minds," as Said says.

A postcolonial reading of territory should begin with the presupposition that territory is not simply the space that is occupied by a nation; nor is it the area demarcated by, or contained within, state boundaries. This is not to deny the existence of territory. Territory clearly exists. Its concrete effects are apparent everywhere. Rather, it is to ask after its production. How is it that territory comes to have an apparently "natural" association with the nation-state? What explains the apparent completeness of the territorial division of the world – as Lenin put it, "the world is completely divided up" – despite the fact that redivisions and redefinitions of space continue? What makes that space I have referred to as "southern Belize" Belizean space? As opposed to, for instance, Guatemalan space? Or Maya space? Or something other entirely?

To answer these questions, we must accept that territory is an effect of the practices that constitute it as such, that is, as the becoming-space that articulates the nation with its state. Territory, in short, is the fundamental spatial ontology of the modern nation-state. Territory is therefore not simply the spatial extent of a nation-state, but the spatial conditions that allow it to be: the spatiality that is required for it to have its natural character (hence, in Braun's terms, territory is "the state's 'nature'").[74] To say this differently, and to stress the iterative quality of this process, we should say that *territory is an effect of territorialization*, where territorialization can be defined as the production of the space of the nation-state. Territorialization is the iterative process whereby states produce the effect of a spatial-ontological separation between its space and the other's. Territorialization is the name for the process of the working-out of the "spatial relations" that make a given state-society ensemble hegemonic.

To the extent that this definition of territorialization is tautological, the tautology lies in the fact that the very meanings of "state," "nation," and "territory" have been interwoven and naturalized in language. It is, therefore, a necessary tautology, one that indicates something profound about the worldliness of the "modern world." The Oxford English Dictionary (OED) defines "territory" as: "the land or country belonging to or under the dominion of a ruler or state."[75] Territory is the state's space. But, then, what is a state, and how does it come to have its space? The state is usually taken to refer to that constellation of institutions and bodies that constitute the government. For instance, the OED defines state as "the body politic as organized for supreme civil rule and government." The state can also refer to "a body of people occupying a *defined territory* and organized under a sovereign government," and even "the territory ruled by a particular sovereign."[76] A nation is "an extensive aggregate of persons, so closely associated with each other by common descent, language, or history, as to form a distinct race or people, *usually organized as a separate political state and occupying a definite territory*."[77] The very language of nations, states, and territory – in English – presupposes this network of self-referential concepts: nations tend towards states; states are defined by their territories; and territory is the space occupied by the nation-state. What is clear through the examination of these terms – in English, at least, but also in the Western tradition that is under examination here – is a theory of the seemingly natural organization of the world into collective units that bind together race, space, and order. The link between state, territory, and nation is further indicated in the twofold meaning of "country," which first referred to an undefined open space, but is now taken to be equivalent to a territorialized nation-state. Today the world is said to be made up of "countries," but the meaning has shifted: the space of "the country" is no longer open. Lenin was correct: the world's space has been completely seized, territorialized. But the story, of course, does not end there.

As we know from the experiences of refugees, borders, and colonialism, territorialization is not simply an abstract or ideological process. Territory must be constantly *materialized* through the myriad socio-spatial practices that successfully bifurcate "us" from "others," in Said's terms – more precisely, that define the nation-state as spatially coherent. This work always involves the unequal manifestation of power in spatial form. Practices of territorialization work not only at the physical margins of a solid "territory" – for instance with boundaries around sovereign space – and not only in cases of border disputes. Territorialization works "internally" to fill in state-space, unevenly.[78] In fact, we could define territorialization by its effect on the production of space "inside" the nation-state as much as anything. The relative strength of its effects could be measured by the evenness and completeness of the sense that a nation-state is spatially bound, settled, and contained.

Territorialization is not a process that only works for or with reference to the state, since it is the spatiality of the *nation* that is so much at stake. Territory as a fixed, concrete spatial formation is never fully present for the subject of nationalism, that is, the subject that is interpellated (recognizes herself to be) within national-territorial space. This is because the subject's entry into the field of language – for example, through the signs that ensure the subject a place, or national space, in the world – is always split.[79] The territorialization of nationhood is dependent on the successful capture (or interpellation) of "a nation" by a given state-society relation. The relation must not only be spatial – it must have the elements of a spatial relation: orientation, directionality, extent, dimension, etc. – but also of a political character. The space must exhibit a substantial correspondence between territory, nation, and the state.[80]

This suggests that there is no necessary relationship between the nation, territory, and the state. There may be some tendencies towards certain configurations, but any hope of defining these would find that they are fraught with the contingent politics and strategies that prevent the completion of the capture and thematization of the world. This opens space for a different way of reading maps. The defining quality of such a reading would be to call into question the nationalism-territorialization relation. We must recognize that the seemingly *natural* association of territory with nation-states presupposes a prior congruence between nationalism, territorialization, and the state. The triad has been constructed through the forms of writing that distinguish modern statecraft – treaties and constitutions – but also through the techniques of representation that constituted colonial governmentality: the census, map, and museum.[81] The naturalness of this triad is exploited when nations or peoples that do not qualify (legally or discursively) to speak as subjects of a territorial state produce maps that produce these effects.

As many geographers have noted, cartography has been essential to the production of the national-territorial identity of modern spatiality. Maps do not merely express spatial patterns. Maps are texts that naturalize certain spatial representations of the world as real, present, and bounded; national maps do so in ways that articulate the naturalness of this worldliness with the nation-state. The map is crucial to the production of hegemony through coordinating land (territory), a people (nation), and governmental order (state). While maps are clearly not the *only* way of writing the state into or on a national space, it is clear that cartography plays a crucial role in naturalizing the experiencing of space in terms of territory, by weaving the state into the hegemonic orientation of spatial existence. Maps produce particular kinds of space in ways that refract the conditions of their own production (including nationalism) but cannot be reduced to ideology.[82] As Gramsci argues, hegemony is never completely captured by the state or any single ensemble; there is always some space for maneuver. The state is itself partly an effect of discursive practices which require a certain play. Moreover, what is at stake in contesting the making and reading of maps is not merely "the symbolic constitution" of mapped space qua territory, but the very *materialization* of territory in both the sense of a "concrete text" (a map in one's hands) and also concrete sociospatial practices (orientation, directionality). Therefore, an inquiry into territorialization may open productive ways to interpret the relations between colonialism, spatiality, and development.

Spacing

Let me briefly recapitulate my argument to this point. As a consequence of the expansion of capitalism through European colonialism, today capitalism qua development shapes politics, subjectivities, and the worldliness of the world. Attempts to critique development through Marxist approaches to development economics and state theory, while essential for attacking the imperial character of capital and the state, are insufficient in themselves to explain the dynamic forms of colonial hegemony as they have been articulated through neoliberal capitalism. Postcolonial theory provides a crucial extension and modification of the Marxist problematic by confronting nationalism and calling into question the forms of knowledge and power constituted by colonialism. But with its focus on the question of the possibility and politics of subaltern *historio*graphy, the postcolonial literature has avoided the question of how the state produces its space – i.e., territorialization – that has always been at the heart of colonial capitalism. I therefore hope to read the political effects of capitalism qua development in the colonial present in a way that simultaneously investigates the spatial-ontological basis of the nation-state, namely territory. In so doing I hope to extend the Marxist and postcolonial theoretical purview to consider the colonialism-development problematic within a stronger philosophical register. The object of research is therefore not development discourses, but capitalism qua development.

Borrowing and shifting a term from Derrida's commentaries on Western philosophy, I refer to this practice of reading as *spacing*.[83] I employ spacing as a name for a mode of reading that calls into question the spatial-ontological thematization of the objects or elements that define a text or discourse. This mode of reading is one that examines the constitution of spaces and highlights the ways spatiality makes possible the work of a text; in other words, it calls into question the ontological productivity of the text's particular spatiality.

One way to understand the implications of this practice is through the postcolonial critique of Foucault. In his archaeologies and genealogies of discursive formations, Foucault demonstrated that spatiality was always already involved with the "historical" production of subjectivity and social power. By demonstrating that "space itself, in the Western experience, has a history,"[84] Foucault's spatial histories of the medical gaze[85] and the prison[86] opened new ways of investigating the constitution of power and subjectivity. This approach helped to inspire the postcolonial project,[87] yet many scholars have criticized his approach on the grounds that Foucault did not consider the conditions of possibility for this category "Western experience" that frames (sometimes in its absence) all of his studies. Ironically, the radical spatialization of history that Foucault's work produces "only reinforces our sense that the place of modernity is to be taken for granted."[88] Though his texts spatialize modernity and raise the question of the constitution of "the West" in a radical way, Foucault does not pursue the becoming-space of European modernity. The colonial and imperial processes fundamental to its condition are left unexamined. For Spivak, this is reason for calling into question Foucault's approach: "Foucault is a brilliant thinker of power-in-spacing, but the awareness of the topographical reinscription of imperialism does not inform his presuppositions. He is taken in by the restricted version of the West produced by that reinscription and thus helps to consolidate its effects."[89] By spacing I mean a reading that questions this "self-contained version of the West" produced through colonialism and capitalism.[90] A postcolonial reading of Foucault takes up the responsibility to map the colonial spaces that make possible the distinction between European and non-European. Development practices have proven especially effective in producing such becoming-spaces of "Europe" and its others.

I consider Spivak's critique of Foucault an exemplar for what I refer to as spacing. Spacing is a mode of reading that calls into question the ground of geographical arguments that rest on spatial empiricism; it challenges geographies – texts that write (of) the world – that begin with location, region, and position, in a way that uncritically takes the texts of the world as they are, without *reading* them. Making room for non-empiricist readings, spacing introduces into the texts that constitute the world space for a productive rearranging – rearrangements that stress the becoming-spaces of particular colonial discourses, and readings that connect distinct points, such as those between imperialism, colonial territorialization, and capitalism qua development.[91]

Spacing makes possible a critical reading of capitalism qua development in the Americas, which are always already related to colonial discourses. In Belize and elsewhere, the ideology of anticolonial nationalist movements coupled development, territory, and nationalism in such a way that indigenous claims – claims to indigeneity, and indigenous land claims – often bring out great ambivalences. Consider figure 0.1: George Price, the leader of the anticolonial, nationalist movement in Belize, is shown teaching a class of young Maya students in a rural village in southern Belize.[92] The photo portrays Price as a caring leader, willing to travel any length to assist in the development of the nation. In the space between Price and the Maya students the drama of postcolonial development unfolds: the development of the nation – of these rural Maya students – depends on the careful trusteeship of the postcolonial state.[93] To read this space, we should notice what frames

Figure 0.1   George Price teaching a class of Maya students, ca. 1964

Source: Ion Media, 2000. *George Price: father of the nation Belize* (ION Media: Belize), no page. Photograph by Norris Hall.

the engagement between Price and the Maya students. On the black-board behind Price, drawn by him for the edification of the Mayas, is a map of the new country. It is divided into political districts and marked with resources ("sugar"; "citrus"; "lobster"). Here development leads the nationalist state to teach the Mayas where they are, where they fit politically, and the nature of their resources. In this image, we can read the hegemony of post-colonial development, which articulates capital-ism and territorialization. Price's map does not recognize the indigeneity of the Mayas; it does not offer multiple interpretations of development or political rights; it does not speak to Maya land rights. It is a map drawn by England and Spain in the course of colonial territorialization and primitive accumulation, carried forward in borrowed terms even as it aims to replace them.

In the course of colonizing Guatemala and Belize, both the Spanish and the British tried to compel the Maya to live in formal communities. While their ways of justifying and attempting this process varied, both were supported by a common belief that the indigenous people would be more manageable, more developable, if congregated into towns (the Spanish colonial term for this was *reducción*). Like Price's map, the colonial practices that have territorialized Maya spaces bind together political identity with *development* and *settlement*. This practice stems from an old affiliation in Western thought between politics, citizenship, and the quality of being spatially fixed, or settled. Colonial power materialized a longstanding alignment of the concepts of city, citizen, and state. It is through this alignment that the centered ontology of the state is justified. It is not only the outsider who is excluded from the *polis*, and therefore politics and citizenship; it is the unsettled subject – the one who threatens nature by moving, by refusing to settle down, by refusing to be a subject of a state.[94] This tradition calls for spacing.

Although spacing is an abstract practice, it is also a purposeful work and an intervention into development politics. What is at stake in spacing Belize is the way we pose the question of the status of the category "Maya" in relation to the modern nation-state. Rather than asking what the status of the "Maya" people is in Belize – are they indigenous? Are they refugees? Are they peasants? – we should ask what constitutes Belize, such that it is a territorialized nation-state in a particular relation vis-à-vis the Maya.[95] This implies a deconstructive reading of the inherited concepts of development, territory, and Mayanism. To confront the latter implies that we reject that part of the Mayanist tradition that seeks to define Maya culture and find, in its essence, cultural resources that are to be preserved and developed. The Maya have been figured in colonial discourse as a scattered, unsettled, fallen race, an epitaph for the failure of civilization. The solution, proposed in different ways across four centuries by Spanish, British, German, and American rulers, has been *settlement* and *development*. In the last few decades, we have entered a new phase of this process, where Maya culture has become figured as a resource *for* development. Before trying to accelerate capitalism qua development on this basis, one should ask how this ensemble of concepts came to have such sway. That is my aim.

Plan of the Book

By now it should be clear that the title of this book is not the name of something that has already happened ("how development was decolonized"), nor something that will take place in this text ("buy this book to learn how development should be decolonized"). Rather, the title names a problem – one that has solicited me, but is not mine: the task of decolonizing capitalism qua development.

Colonial development practices are ubiquitous today, but we often fail to name them and criticize them as such. It is as though they are so pervasive that they become thematically, and therefore politically, impossible to grasp. But from Belize to Iraq, we must recognize that it is through the work of imperial capitalism qua development that the modern world has been constituted as such.[96] Following Timothy Mitchell, I argue that what is at stake in colonialism and development is the very constitution of modernity as the dominant mode of enframing the world. My reading of the colonization and development of southern Belize is not, therefore, an exercise in offering a social history of a place. Rather, I aim to call into question the enframing of development and Belize that have made it possible to think of these things as having a proper, empirical historical geography – one that solicits neoliberal development *and also* silences Maya resistance to these practices because their geographies do not match ours. This is what I mean by "spacing Belize."

So this book is not about the geography of Belize, but rather an intervention into what Edward Said called "the struggle over geography."[97] As the etymology of "geography" suggests, geographers should aspire to world-writing. The work of a geographer should question and unsettle the presuppositions about space and geography that underlie the hegemonies of an unjust world.[98] I understand geography here in an anti-disciplinary sense, as a hermeneutic practice that aims at interpreting the practices that constitute the worldliness of the world.

I attempt to do so in Part I by reading texts in the subaltern history of southern Belize (chapter 1) and by examining the discourse of the Maya farm system (chapter 2) and its place in the broader archaeology of Mayanism (chapter 3). Part II examines aporias of development by investigating the works of A. C. S. Wright, a colonial soil scientist whose works were instrumental to transforming colonial discourse into a discipline of development (chapter 4), of two development projects that aimed at "settling" Maya agriculture (chapter 5), and of a counter-mapping project, the *Maya Atlas* (chapter 6). This may suggest a chronological development, and so I repeat, the colonial period has not ended. No. Since my argument is that colonialism solicited development as a form of power appropriate to winning hegemony for capitalism, there are no clear lines separating a colonial past from the development present.

Notes

1   World Bank (2003). Downloaded from www.worldbank.org, July 2003. The statistics that follow are also from this source. GNI calculated by the Atlas method. Africa and Latin America data are from 2002.
2   For a review of the data on global inequality, see Wade (2004). Any meaningful definition of poverty must consider relatively inequality.
3   Gramsci (1992, 1996).
4   World Bank (2003). Belize's GNI per capita is comparable to Botswana ($2,980) and Brazil ($2,850). Belize compares unfavorably with other countries in its income group in terms of education and level of industrialization, but very favorably with respect to health care and environmental conditions. For instance, the life expectancy in Belize (73 years) is seven years longer than the world average, and the under-five mortality rate is only half of the world average. Yet these relatively strong indicators do not apply to the rural south.
5   Data calculated from Belize Census 2000. "Maya communities" are defined here as census communities of more than 30 people where more than two-thirds speak a Maya language (mainly Q'eqchi' and Mopan, but also occasionally Yukatek) as their first language.
6   The Bank's views matter intensely because it is at the center of a cluster of institutions that shape development discourse on a planetary scale. This is not the place to review all the criticisms of the Bank (but see Caufield 1996). Most of the Bank's loans support large projects – particularly dams and energy projects – that produce modest returns. Loans must be repaid to the Bank regardless of whether the projects create economic growth or improve the quality of life for the poor; according to the Bank's own estimates, about half of the projects they have funded have failed in their own terms. This contributes in a minor way to the steady net flow of capital from the Third World to the core industrial economies. This flow has not only led to a steady growth in the capital stock of the Bank itself, but allows the Bank to employ "some 10,000 development professionals from nearly every country in the world" (World Bank 2003, cited at www.worldbank.org, July 2003).
7   The following sectors have been privatized since the early 1990s: water, energy, port and dock services, telecommunications, and airport services. Hundreds of government positions were eliminated in a sweeping "retrenchment" before Christmas in 1995.
8   These concessions to the forests of Toledo attracted capital from Mexico, the US, and China, to expand logging and sawmill operations (see chapter 4).
9   Like Casteñeda in his study of Mayanism and tourism in the Yucatan, this book "must necessarily suspend designation of the Maya and resist the seduction to define who or what is Maya, since it is precisely my task to critique and contest how this has been done" (Casteñeda 1996: 13). See also Montejo (2005).

10 As Julian Cho, leader of the movement between 1995 and 1998, explained: "Indigenous people have been very passive; we have always been taken for granted, that we should sit down and we should not voice our opinions.... [W]e are one people and we are going to speak as one people" (1995).

11 These maps were published in TMCC and TAA (1997).

12 I am involved in the struggle to materialize the IACHR ruling as a member of the advisory council of the Julian Cho Society (JCS), an indigenous-rights NGO in southern Belize; as an "expert adviser" to lawyers collaborating with Maya communities to win a case against the Government in the Supreme Court of Belize; as an assistant to an ongoing counter-mapping project; and in other roles (see Wainwright 2007).

13 This has been a matter taken up in the literature on political ecology. This book could be read as political ecology if we define this field as an anti-disciplinary project aimed at calling into question the nature of politics and the politics of nature (see Wainwright 2005).

14 I consider Mayanism in Part I and ethnography in chapter 5. Apart from the critique, I see no need for more ethnographic studies of the Mayas of Belize, since that work has been done (for the best of it, see Wilk 1997). There is a massive literature on Maya archaeology and history. According to a search of dissertations, between 1996 and 2003 there were 58 theses written with the keywords "Belize" and "Maya" in the title or abstract. Of these, 45 (78 percent) are archaeological; six concern forestry or political ecology (8 percent); five are ethnographies of living Mayas (6 percent). Similar statistics could be compiled for Guatemala and southern Mexico. The Maya are perhaps the most intensely studied ethnos in the world.

15 If we define "success" narrowly – in terms of articulated, growing, economies – we could say that colonialism did not create the conditions for successful capitalist development. Colonialism did sometimes create conditions for rapid economic growth, such as in Taiwan and South Korea, but these are exceptional cases, not least for the subsidies received from the US (market access, capital investment), which had an interest in seeing these economies grow rapidly (see Chibber 2006).

16 For four distinct views from the Marxist left on the question of imperialism and global capitalism, see Amin (1989); Hardt and Negri (2000); Smith (2003); Karatani (2005).

17 The experiment with neoliberalism in Belize has produced effects that are all too well-known elsewhere: privatization of state properties, class stratification, and uneven development. For a study of the effects of neoliberal policies in Belize, see Gabb (1992); for a theoretical critique of structural adjustment, see Carmody (2001). Unlike Zimbabwe (Carmody's case study), Belize had little industry to lose through neoliberalism, but structural adjustment has encouraged disinvestment in agriculture and what industry exists.

18 "Property" is of course an extraordinarily complex concept. The sign "property" weaves together belonging, responsibility, ordering force, and correctness. Development is properly of the territorial capitalist state.

19 See Mitchell (2003).

20 As with the other nation-states of the Americas, Belize was forged through slavery, the decimation of indigenous peoples, primitive accumulation, ecological destruction, forced displacements, and territorialization – the original, foundational violence that cleared the way for contemporary forms of hegemony. For subaltern histories of these processes, see Bolland (1977, 1997); Shoman (1995).

21 Williams (1983: 184).

22 Ibid., p. 188.

23 These words stem from an earlier English form, "disvelop," closely related to the modern Italian *viluppare*, meaning "to enwrap, to bundle" (OED, Vol. 4, p. 562). This sense became coupled with the Darwinian notion of "evolution" in the nineteenth century, and today this meaning of "to develop" is almost synonymous with "to evolve." Here too the parallel with nature holds (see Williams 1985: 189).

24 Aristotle (1941: *Physics*, p. 238); my italics.

25 Howard (1992: 79). In the Introduction to his *Lectures on the History of Philosophy* Hegel argues that "the history of philosophy is the history of...reason"; tracing this history, Hegel writes, requires defining three elements: thought, concept, and idea. Hegel characterizes the movement between these three elements as "development": "The Idea as development must first make itself into what it is. For the Understanding, this seems to be a contradiction, but the essence of philosophy consists precisely in resolving the contradictions of the Understanding" (1985: 71). This gives rise to the distinction between potential and existence. As with Aristotle, Hegel uses the metaphor of the development of the seed into a plant for the former.

26 I am paraphrasing Cowen and Shenton's (1996) thesis here. Cowen and Shenton name the coalition of the immanent process with an intention to develop as a "doctrine of development." While their broad analytic history of development doctrines is important and useful, the task remains to specify the relation between colonialism, capitalism, and development.

27 Aristotle (1941: *Physics* II: 1, p. 26).

28 For instance, the value of Adam Smith's argument about the so-called "invisible hand" was to show that the free movement of capital and commodities, guided by the pursuit of self-interest by rational individuals, leads national communities towards good ends. One could write a genealogy of the processes that have produced this attachment (in which Smith's texts would be one accomplice).

29 On the "worldliness of the world" see Heidegger (1996: section I.III).

30 Adorno (1998 [1962]: 143). On this "historical identity," see Chatterjee (2001: 169) and below.

31 On the political economy school of development geography, see Samatar (1989, 1999); Glassman (2004); Carmody (2001).

32 See especially Ferguson (1990); Escobar (1994); Crush (1995); and Rist (1997). Post-development is comprised of a diverse collection readings of development practices, all of which suggest some poststructural influences, particularly Foucault. But these influences are applied unevenly and often in a simplified, functionalist way by post-development writers.

33 Ferguson (1990: xvi).

34 Ferguson concludes that development is an "anti-politics machine." But we must reject this formula. Development is of course political, albeit in a way that does not necessarily produce liberal democratic spaces, and very often today development projects and discourses frankly admit this. Indeed, few things are as productive of political spaces and subjects as capitalist development. The question, then, is not whether development stifles politics, but rather how and why certain political surfaces are produced and contested under the hegemony of development.

35 Cf. Watts (1993, 1999, 2001a); Lehmann (1997); Corbridge (1998); Blaikie (2000); Pieterse (2000).

36 Lehmann (1997: 572).

37 Escobar (1994: 4).

38 Ibid.

39 Gidwani (2002: 5–6). Following Ajay Skaria, Gidwani proposes that development be thought as "a placeholder concept that denotes regulatory ideas about a "better life'" that varies in space and time and always carries "multiple accents." At least in his 2005 essay, Gidwani does not explain how development has come to be not a placeholder, but a synonym, for capitalism.

40 For Watts' critique of post-development, see Watts (1999).

41 Watts (2001a).

42 See Derrida (1978 [1966]); Ismail (2006). I discuss ethnography in chapters 3 and 5.

43 Some post-development writers have suggested that the Age of Empire gave way to the Age of Development (Escobar 1994), but late nineteenth-century European colonialism was typically justified in terms of "development." At its best, post-development draws our attention to the discursive effects of capitalism qua development. But this makes it something other than "post-development" – "cultural studies of development" is a more appropriate name – since we are after all still living in a world defined largely through capitalism qua development.

44 Cowen and Shenton define trusteeship as "the intent which is expressed, by one source of agency, to develop the capacities of another" (1996: x).

45 "Cannot not desire" is one of Spivak's formulas for the deconstructive scenario.

46 Spivak (1999: 426–7).

47   Spivak (1994: 52).

48   Derrida (1995: 61).

49   I emphasize "and" because capitalist states do not choose between imperial rule and hegemony; empire and hegemony are not mutually exclusive (see chapter 3).

50   See especially Spivak (1994); Gupta (1998); Gidwani (2002, forthcoming); Mitchell (2003); Kothari (2005); Moore (2005). On development geography research that draws upon postcolonialism, see Sidaway (2000); Radcliffe (2005); Gidwani (forthcoming).

51   Cf. Shohat (1992); McClintock (1992); Mowitt (2005). Part of the challenge is that postcolonialism resists reduction as a framework, which is the organization of knowledge that development studies typically seeks.

52   Ismail (2006: xvi); also Mowitt (2005: xxvi–xxix).

53   This Heideggerian argument has been advanced by Mitchell, who posits that modernity is constituted as "an effect we recognize as reality by organizing the world endlessly to represent it" (2000) In this view, what becomes extended through colonialism is a set of practices for representing the world: "Representation does not refer here simply to the making of images or meanings. It refers to forms of social practice that set up in the social architecture and lived experience of the world what seems to be an absolute distinction between image (or meaning, or structure) and reality, and thus a distinctive imagination of the real" (2000: 17).

54   One of the reasons that postcolonialism is not simply a project of subaltern historical research is that the practices that constitute the present have shaped the very conception *of* history (cf. Lalu 2005). To put this differently, postcolonialism concerns the colonial *present*. On postcolonialism and subaltern historiography, see Chaturvedi (2000).

55   See Ismail (2006).

56   Fanon (1998: 235), my italics.

57   Chatterjee (2001: 168).

58   The expression "nationalism qua nationalism" is the key to this argument. What could nationalism be *except* nationalism? Can we imagine a politics that moves in the space of nationalism that could challenge the legitimacy of Reason and capital – in a way that "mere nationalism" cannot? The thrust of Chatterjee's argument (which opposes the distinction between "good" and "bad" nationalisms) suggests that the answer is no. This is part of what distinguishes postcolonialism from anticolonialism. Postcolonialism struggles against colonialism and nationalism, which only "produces exclusivist notions of community and might be understood as a structure in dominance" (Ismail 2006). In this study I use ethnos in lieu of race and nation not simply to draw out the connection with ethnography, but because the ambiguity between ethnos' "positive" troping (as healthy nationalism) and "negative" (ugly racism) in the tradition of liberal thought is problematized by postcolonialism.

59 Chatterjee (2001: 169). Chatterjee comes close to Heidegger's position here. I see no justification for reducing this dynamic to a question of science and technology. Finance is fundamental.

60 See Spivak (1988, 1999). Spivak's original argument was an intervention in debates of the Subaltern Studies group on the question of the very possibility of reclaiming subaltern histories (cf. Guha 1982, in Chaturvedi 2000), but it applies equally to subaltern geographies.

61 Lalu (2000: 68).

62 Casteñeda (1996: 13).

63 Ibid.

64 On pan-Mayanism, see Casteñeda (1996); Warren (1998, 2001); Montejo (2005). On the construction of "the Maya" and "Maya tradition," see Hervik (1999) and Greene (2003). This subsection of the Introduction uses language previously published in Wainwright (2005). I thank Pion Limited (London) for permission to use these revised extracts.

65 Spivak (1999: 408–9).

66 Cf. Chakrabarty (2000).

67 Heidegger (1996). One illustration of the power of this way of viewing the world comes from the World Bank. The (wholly appropriate) sign for the Bank is the world itself: a portrait of the world made up of latitude and longitude lines, signs of abstract space *par excellence*.

68 Gramsci (1957: 198–9). Gramsci argues that the debate about the "so-called reality of the real world" demonstrates the power of Judeo-Christian theology on "common sense": the apparent "realness" of the "real world" is self-evident, since it is God's creation. To doubt its realness – or even question the relationship between the real and the thought of the real – is to doubt not only the Creation story but God's existence, since the grounding myth of Judeo-Christian metaphysics is that God created the world (see the discussion of Las Casas in chapter 3). Gramsci's argument that the "problem of the reality of the real world" cannot be posed within popular philosophical terms closely parallels Heidegger's argument, made at roughly the same time: "With respect to the problem of an ontological analysis of the worldliness of the world, traditional ontology is at a dead-end – if it sees the problem at all" (1996: 61).

69 Lefebvre (2002: 84). The other two moments that Lefebvre specifies are "the production of a social space" and "a mental space." A postcolonial reading of territorialization should reject Lefebvre's trifurcation of mental, social, and physical, and also his inference that production of the national territory, by virtue of its concreteness, is therefore natural. This assumption recapitulates the Western metaphysical propensity to equate materiality, nature, and substantiality.

70 Cf. Agnew and Corbridge (1995).

71 Johnston (1995: 620).

72 Brenner argues that the Westphalian model of the nation-state-territory coupling did not become hegemonic until the twentieth century, when "the notions of state, society, economy, culture, and community... had come to presuppose this territorialization of social relations within a parcelized, fixed, and essentially timeless geographical space" (1999: 47).

73 See Said (1979: passim).

74 Braun (2000: 28).

75 OED, Vol. 17, p. 819. While the etymology is said to be "unsettled," it seems that territory shares a root with *terrere*, to frighten, from which we have "terror." Territory would thus be something like a space "from which people are warned off" (ibid.). Territorialization is the process whereby the nation-state warns off the other. In fact the OED defines the verb "to territorialize" (from which territorialization is derived) as "to place upon a territorial basis; to associate with or restrict to a particular territory or district."

76 Ibid.

77 OED, Vol. 10, p. 231. Italics mine.

78 For another treatment of internal territorialization, see Vandergeest and Peluso (2001).

79 Althusser's scene of interpellation is already spatial: the policeman's hail says "hey, you *there*"; the friend's saying "It's me" occurs with a knock on a familiar door; etc. These scenarios not only presuppose an ideological environment in which interpellation can produce subjects, but a particular spatiality in which the subject exists. Interpellation, for Althusser, produces spatial subjects.

80 Kojin Karatani proposed a similar problematic with the central trinity of his transcritical Marxism: nation-state-capital (2005: 13–16). In my view the key trinity is (nation–state)—capital—territory. It is the spatial, especially territorial relations of power that makes particular nation-state-capital couplings possible.

81 Anderson (1983). "Development" has played an equally critical role in binding the modern nation-state. Development not only refers to the evolution of the nation under the guidance of the state; it solicits the very thought of such evolutionary improvement as a way of figuring the temporality of the nation-state. Capitalism qua development calls for trusteeship on the terrain of the nation-state-territory complex.

82 Doing so would put one in the position of searching out the ways that ideology has "distorted" the "correct" representation of the world in a map.

83 I stress *shifting* because my use here cannot but flatten Derrida's complex uses of the term *espacement*, or "spacing," in many of his most important works from the 1960s (see "Différance"; "Ousia and gramme"; "Freud and the scene of writing"; see also *Of grammatology*, and the interviews in *Positions*). Spacing is not a neologism in English. It is the verbal substantive of the verb "to space." The OED defines spacing as "the action of the verb

[to space], in various senses, or the result of this action" (OED Vol. 16, p. 93).

84   Foucault (1998: 176).

85   Foucault (1973). Foucault finds that in nineteenth-century France "a quite new form, virtually unknown in the eighteenth century, of institutional spatialization of disease, makes its appearance. The medicine of spaces disappears" (20). See also Foucault (2003 [1976]: 103).

86   Foucault (1979).

87   Cf. Said (1979). In *Culture and imperialism* Said says of Foucault and Raymond Williams: "I am in considerable sympathy with the genealogical discoveries of these two formidable scholars, and greatly indebted to them. Yet for both the imperial experience is quite irrelevant, a theoretical oversight that is the norm in Western cultural and scientific disciplines" (1993: 41).

88   Mitchell (2000: 5). Mitchell's work attends to this problem by drawing from Heidegger and Derrida to examine colonial power. His work suggests that "it was in the building of slave-factories in Martinique, prisons in the Crimea, and schools in Calcutta that the decisive nature of the distinction between European and non-European was fixed" (3). To this list we may add the Indian reservations of British Honduras (see chapter 1).

89   Spivak, cited in Mitchell (2000: 29).

90   These are Spivak's words, but the tone is Lenin's. The revised version of this essay in *Critique of postcolonial reason* concludes by quoting Lenin to the effect that "Capitalism has been transformed into imperialism" (1999: 311).

91   Consider the ruins of Lubaantun near the village of San Pedro Columbia in southern Belize. Geographers have had a strong language for talking about where Lubaantun *is*, at least since 1928, when Lubantuun was located by surveying techniques: "[The] character of the expedition…was partly geographical, partly archaeological. In its origin it was archaeological. That is, the British Museum had the opportunity of sending out an expedition to investigate something of the archaeology of British Honduras, and the Royal Geographical Society took the opportunity of sending Mr. Laws attached the expedition to make such geographical exploration as was possible. That was, I think, mutually advantageous. It was helpful to our party to have Mr. Laws with it. He fixed the position of Lubaantun and, I hope, has established in your minds the fact that it does really exist, and he was also able to contribute something to the geography of the surrounding country" (Laws 1928: 236). What does it mean that Laws "fixed" the position of Lubaantun? In Kenyon's text, Lubaantun is fixed by Laws because he correctly identified its position in a universal and objective spatial grid. The space in which it is identified belongs to the becoming-space of presencing, is the spatiality of the Cartesian coordinate system, although this particular worlding of the world is unquestionably the

hegemonic one today (see Heidegger 1996). That there are other ways of worlding the world is fundamental to the concept of *spacing*. Fixing, or settling, names the operations of positioning the Cartesian subject *in* a world that already "is" spatial – which the subject merely fills with a body.

From this example it may seem that spacing is a mode of reading that applies only to visual and "spatial" texts. Yet consider Glenn Gould's interpretation of the music of J. S. Bach. Gould's recordings and critical writings offer a radical mode of reading and performing Bach, a mode that Edward Said has described as "contrapuntal." I would emphasize the way that Gould brings out the contrapuntal elements of Bach – i.e., plays the notes "out of time." In Gould's approach we may hear spacing being put to work. Gould's favored practices – total abstention from the sustain pedal; the heightened shifts in tempo; the arpeggiation of chords – increase the severity and contrapuntal sound of his playing. Through this productive "misreading" of Bach's scores, Gould brings out tones, combinations, and potential lines that would not otherwise exist. Although Gould's is a practiced reading, it is more a habit than a conscious technique. Consider Gould's comment from a 1976 interview with Bruno Monsaingeon, inserting ourselves when they are discussing Gould's practice of eliminating sforzandos. Monsaingeon suggest that sforzandos "represent a disruption of counterpoint," and that Gould's relentless "search for counterpoint compels [him] to change or to 'correct' voice leading." Gould affirms: "I very often arpeggiate chords which are written conventionally... and I read quite often the gentlemen of the press assume this is some sort of parlor-music mannerism. In fact, it's quite the opposite. It may or may not be justified, but the habit originates in a desire to keep the contrapuntal spirit alive, to emphasize every possible connection between linear events.... The nature of the contrapuntal experience is that every note has to have a past and a future on the horizontal plane" (Gould 1984: 36). Gould's practice (or "habit" as he says) is a powerful illustration of spacing. By "emphasiz[ing] every possible connection between linear events," space can be thought as something other than the inert and discrete container of being as presence. Agamben says as much when he writes of Gould: "even though every pianist necessarily has the potential to play and the potential to not-play, Glenn Gould is, however, the only one who can *not* not-play, and, directing his potentiality not only to the act but to his own impotence, he plays, so to speak, *with his potential to not-play*" (Agamben 2005: 36, my italics).

92 Anon. (2000).

93 I use the terms "post-colonial," "postcolonial," and "anticolonial" to refer to distinct phenomena. Post-colonial is always used as an adjective to describe a nation-state in the time after the end of formal colonial rule. For example, "the post-colonial Belizean state" is the one that governs Belize after September 1981. This term is not to be confused or conflated

with postcolonial, which refers to a project of challenging and undoing colonial thought and politics. Postcolonialism troubles the notion of the post-colonial insofar as it undermines the thought that states or time periods that occur after the end of colonialism have broken from colonialism. I know of no way around this terminological difficulty.

94　See Agamben (1998). Notably, in the famous passage from the first book of *Politics* where Aristotle defines man as a "political animal," Aristotle argues that the category "citizen" is reserved for those who live in the *polis*: "The final association, formed of several villages, is the city or state ... [which is] a perfectly natural form of association, as the earlier associations from which it sprang were natural. ... It follows that the state belongs to a class of objects which exist in nature, and that man is by nature a political animal; it is his nature to live in a state" (1979: 28–9). The Western political tradition since Aristotle has carried forward, in some fashion, this affiliation between settlement as politico-spatial fixing (to live *in a state*, i.e., be territorial) and state-citizen agreement (contract as natural form of association).

95　This implies a critique of Maya ethnography, the literature that has been most responsible for specifying who the Maya are and how they fit "in Belize."

96　I do not mean "modern" in the sense of "contemporary," but as that which is made by the worlding of a vulgar time that posits temporality as a common, evolutionary horizon of human development – in which "the West" is ontologically separated from, and ahead of, "the rest." See Mitchell (1988, 2000).

97　Said (1993: 7).

98　No climate will be found in this book, nor descriptions of soils, roads, population, or cultural groups. I know these might be confused with the geography of Belize. For such data, see: www.cia.gov/cia/publications/fact-book/geos/bh.html#/Intro. What does it mean for geographers that the CIA "does geography" (as this practice is widely understood) as well as anyone? There is still much work ahead in remaking geography as something other than an imperial and empiricist science.

Part I

Colonizing the Maya

1

Unsettling the Colonial Geographies of Southern Belize

The history of the subaltern classes is necessarily fragmented and episodic; in the activity of these classes there is a tendency toward unification, albeit in provisional stages, but this is the least conspicuous aspect, and it manifests itself only when victory is secured. Subaltern classes are subject to the initiatives of the dominant class, even when they rebel; they are in a state of anxious defense. Every trace of autonomous initiative is therefore of inestimable value.

Antonio Gramsci, 1930[1]

In 1918, John Taylor, former warden of the colonial prison and then District Commissioner of the relatively new Toledo District of southern British Honduras (as Belize was then known), typed these words regarding the "Indian question" to his superiors:

Much could be written on the subject of [the] Indian population, – they are interesting, – I for one would like to see an improvement in this (to my mind) – fast decaying Race, – they are, especially youngsters Bright and Quick to learn, – and although these kiddies when in school appear to me studious, and seem to enjoy it, yet they much prefer – the Boys – to shoulder a Machete and strut off with Father to the Milpa. Any improvement in their mode of living, or agricultural methods, can only be brought about by others foreign to them, either by example or inducement, – Force beyond a certain point will not do.[2]

By the time Taylor typed these lines in his Punta Gorda office, British policies and positions towards the Mayas in Belize were well established.

The Mayas were to live within Indian reservations, supervised by alcaldes who were incorporated into the colonial state. Alcaldes were responsible for collecting land taxes and enforcing colonial law. Living in settled villages and attending school were mandatory (though education was left to the Catholic church). To sum up British policy in a word – one that appears frequently in the colonial archive – the Maya were to be *settled*.

Like many of the colonial texts of the era, Taylor's comments are replete with the essentialist tropes about those he was sent to govern. Unelected and racist he may have been, John Taylor was the sole legal authority and magistrate of the Toledo District for more than ten years. Even the minor texts and statements of such colonial officials had concrete effects. For instance, Taylor's claim that the Mayas are by nature "childlike" and "fast decaying" calls for the intervention of a paternal state to act as trustee for the Maya, frames the Maya as racialized subjects, and solicits the state as a development institution (to end the decay). We must pay attention to such texts, since it is by analyzing colonial discourse that we may come to understand colonial hegemony, understood in Gramsci's sense as the forms of moral and intellectual leadership that ramify through cultural practices and sustain unequal social relations.[3] Social groups compete for hegemony in order to consolidate projects that facilitate capital accumulation to their advantage. Taylor's argument that "force beyond a certain point will not do" stands as evidence of Gramsci's insight. Taylor was keenly attuned to the problem of calibrating the use of force: hence his qualification that "Force *beyond a certain point* will not do." Force is needed, but only to a point. How is the point defined? Not ethically, but in terms of efficiency. Beyond a certain point, the costs and difficulties of coercion – including the likelihood of provoking resistance – outweigh the benefits of the desired change. Hegemony names this "certain point." It is the constellation of forces that "will do" – that produces consent. The British colonial state was authoritarian, but it could not maintain overwhelming force or rule through explicit coercion. The colonial state there – like colonial power generally – needs consenting subjects and territorialized spaces. This chapter aims to discern these forms of colonial hegemony, subjects, and territorial spaces.

The Colonization of Southern Belize

Two narratives orient the historiography of Belize: one tells of the heroic victories of the British settlers in their conflicts with Spain and the Mayas; the other relates the unfolding discovery of the nation's geography through scientific exploration and description. Both narratives are geographically deterministic insofar as they suggest that Belize's history is a function of its geographical location, climate, and resources.[4] Moreover, they frame Belize's history teleologically. Thus colonialism and capitalism are quite literally *naturalized*. Consider Lucas's summary of Belize's history:

> From an historical point of view, British Honduras is a very interesting instance of the evolution of a colony. It began with private adventurers, who held their own in spite of a strong foreign power [Spain] and whose success practically obliged their own government to afford them some measure of recognition and protection. It originated with trade, trade begat settlement, and settlement brought about in fullness of time a colony.[5]

Framed in this way, the historical development of Belize is reduced to the mechanistic unfolding of colonial capitalism, a natural development. Insofar as agency is attributed to historical actors, it is located with the British settlers – and, to a lesser degree, their Spanish competitors. Contingencies in the flow of this history are attributed to heroic Europeans. This sort of teleological historiography justifies colonialism and marginalizes non-European subaltern voices.

British buccaneers from Jamaica settled at points along the Central American coast in the 1650s, including the delta of the Belize River, where they began cutting logwood for export to England. Two centuries later, this outpost of colonial capitalism become the capital of the British colony of British Honduras.[6] The Spanish claim, legally recognized by England until 1798, delayed the development of state institutions.[7] The territorial status of the area now known as southern Belize was especially unclear, since treaties between England and Spain only covered the land as far south as the Sibun River near the center of the country. Although contact between Mayas and Europeans in southern Belize may have occurred as early as the 1520s, when Cortés marched southward through the area now known as the Guatemalan *Verapaz*, southern Belize had not been settled by the Spanish; uncolonized, it remained a contested space, claimed by two European states yet inhabited by Manche Chol and Mopan Mayas.[8]

British efforts to colonize southern Belize did not commence until the colonial state was at war with Mayas in other regions of the colony. Although Maya people lived in southern Belize before the 1880s, the state had no contact with them.[9] The area south of the Sibun River was essentially *terra incognita* to the colonial state:

> *The Southern portions of our territory have never been explored*, and according to the Crown Surveyor they contain inhabitants who, he believes, have never yet been seen by European or creole. The rivers south of the Sibun have their source in the mountains whose line of water-shed forms the division between ourselves and Vera Paz. Adown these streams...Mr. Faber has seen floating, rough wooden bowls and other implements which testify to the existence of some inhabitants utterly unknown to us.[10]

The search for mahogany by European loggers drew capitalists, and the liminal colonial state behind them, towards the source of these wooden bowls. Between the 1840s and 1880s, logging crews came into occasional contact with Mayas in southern Belize, but contact between the state and the Mayas was infrequent and did not decisively shape state policy toward the Maya (there were as yet no state institutions in the south).[11]

The event that eventually caused the British to recognize the existence of Mayas in southern Belize was the flight of landless peasants from the Alta Verapaz into the lands around the present-day communities of Pueblo Viejo and Aguacate in the 1870s and 1880s. During this period, Guatemalan land and labor laws were changed to facilitate the expansion of capitalist agriculture. The effect of these policies was felt immediately in the Alta Verapaz (a region inhabited mainly by Q'eqchi'- and Mopan-speaking Mayas) through the explosive growth of coffee plantations. Between 1858 and 1862 alone, 75 coffee fincas were created on lands that had been held in common by Q'eqchi' communities around Cobán and Carchá. By the 1880s, thousands of Mayas had fled the Verapazes to the north, into the Peten, and to the lands along the rivers in the east. Exile denied labor to the coffee estates.[12] The existing Maya communities in southern Belize – a heterogeneous group of Q'eqchi', Mopan, and Manche Chol-speaking Mayas – grew with the influx of migrants. By the 1880s, ~1,500 Maya people were living in what is today the Toledo District – a political space that did not yet exist.[13]

The accumulation strategy of colonial Belize had two major elements: the export of forestry products to the US and England and the import of food and manufactured goods from England for consumption in Belize. These paired movements produced a regular flow of capital to British capitalists. Capital generated from exporting primary commodities to the US went to British manufacturing capitalists, since most of the capital generated through exports was used to import commercial goods from England. The main benefactors within the colony were the large shipping houses in Belize City that imported and sold food and manufactured goods (particularly cloth and clothing). What made this accumulation strategy especially profitable for British capital was the fact that the two most important factors of production, land and labor, were derived through primitive accumulation.[14] From the perspective of British capital, the effectiveness of this strategy can be measured by the fact that the forests of Belize were almost entirely cut over twice before any substantive buildings, roads, or state institutions – apart from taxation and policing functions – were built in the colony. Throughout the nineteenth century, unprocessed mahogany logs accounted for more than half of the total value of exports from the colony. These exports peaked in the mid-1840s, a period when the European railway boom led to consistent demand for mahogany. This mahogany boom was followed by a bust triggered by overproduction: within the colony, mahogany had been overcut; and internationally, prices declined as exports to Europe and the US increased from other regions.

The boom and bust cycles of overcutting and land speculation occurred in the south later than the north and west. Maps from the surveyor's department indicate a mahogany boom throughout southern Belize in the 1860s.[15] Figure 1.1 shows the location of 22 concessions to log mahogany on the five southernmost rivers in Belize. At this time the land along the Golden Stream, Rio Grande, and Moho rivers was divided into logging concessions.[16] These maps suggest that the extent of the logging did not go far inland, at least before 1861. (Figure 1.2 shows an overlay of the 1861 map, transformed with a GIS using the rivers as control points, onto a contemporary map of southern Belize.) The forestry concessions led to the first concerted attempt to parcelize private estates and created southern Belize's first real estate market. By 1868, almost all of the land in a 10-mile strip near the coast, between the Temax and Deep rivers, was privately held (see figure 1.3). Many of the boundary lines of these new estate properties correspond to the boundaries of the earlier logging concessions.[17]

There is scant evidence that any of the capital generated by slavery and logging in the nineteenth century was reinvested in productive activities within the colony.[18] Surpluses were typically invested in land speculation, which contributed to Belize's longstanding land monopoly. In 1787 twelve settlers "owned almost all of the land" in British Honduras.[19] Initially these monopolies did not cover the land in the Toledo

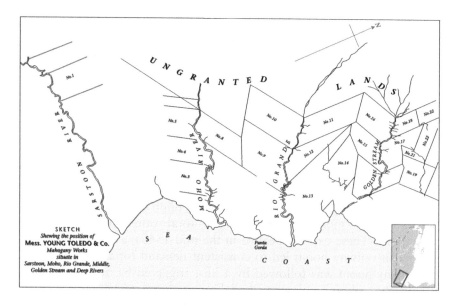

Figure 1.1   Sketch map showing logging concessions, 1861

Cartographer: Eric Leinberger, 2005

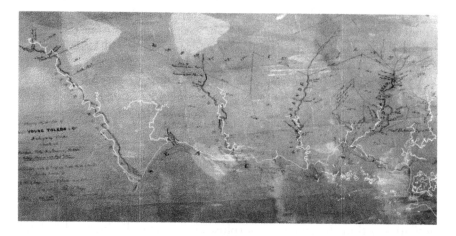

Figure 1.2   Sketch map showing logging concessions, 1861 (transformed)

Cartographer: Eric Leinberger, 2005

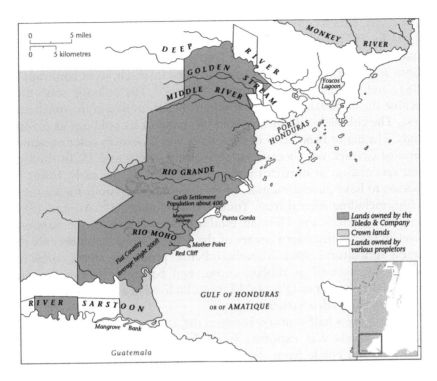

Figure 1.3 Map of Crown and private lands, 1868 (transformed)

Cartographer: Eric Leinberger, 2005

District (which did not yet exist as such) since the lands south of the Sibun River lay beyond the treaties with Spain. Most of the land that remained was not converted into private estates until the mid-nineteenth century. Multinational firms that held sufficient capital to take advantage of the decline of mahogany in the 1850s benefited by accumulating land titles from smaller settler-owned firms. By 1881 one company, the Belize Estate and Produce Company (BEC), owned over a million acres of land – roughly half of all the private land in the colony.[20] Unlike the relatively independent, small settler companies, the BEC used their power to lobby the British government to convert their logging concessions into titled property, thereby gaining power relative to the settlers and the colonial state.

One of the first companies to gain ground in southern Belize was the Young, Toledo, and Co., founded around 1839. The company accumulated extensive properties after the 1854 passage of the first Land

Titles Acts, which converted mahogany concessions into titled property rights. By 1871 the company had acquired more than a million acres of land, including most of the private land available in southern Belize. The boom was brief. The company overextended its reach, went bankrupt in 1881, and lost its properties to the Crown.[21] The colonial state thus became the main landowner in southern Belize, unlike in the north and west. The colonial state in turn sold off many of its holdings as private lands. The major benefactor of these sales was a German colonist named Bernard Cramer, who became wealthy in the 1860s and 1870s through land speculation in northern and central Belize. In the early 1880s he appears to have purchased most of the logging concessions for southern Belize, including several from Young, Toledo, and Co.[22] Around 1891, Bernard Cramer's son Herman established an estate on the Sarstoon River in the southwestern corner of the colony, which soon became the largest agricultural estate in southern Belize, producing all the coffee for the colony, as well as rubber, cocoa, and bananas. (Cramer's estate employed a number of Q'eqchi' Mayas who lived in and around a village known as San Pedro Sarstoon.)

For roughly a half-century, between the 1880s and the 1930s, a wide range of goods was exported from the Toledo District: mahogany, logwood, and chicle from the forests; sugar and rum from the sugar estates along the coast; bananas in the Rio Grande and Monkey River watersheds; and cocoa, rubber, coconuts, copra, plantains, and coffee from the estates and Maya communities in the interior. The forests and laborers of southern Belize also produced chicle, cacao, and cohune nut oil for export to the United States until the Depression, when export prices crashed.[23] Yet the extensive production of southern Belize during the colonial period generated almost no local capital accumulation. The colonial state did little to counteract this unevenness. Almost no tax revenue was collected from the land and timber monopolies, and there is no evidence in the colonial record or on the landscape of any investment in rural Toledo.

Even into the twentieth century, state institutions remained weak in the Toledo District, partly as a result of the area's reputation as the most unhealthy and remote in the colony. Although state institutions in southern Belize were small, the local state consistently ran deficits, especially during periods of economic growth. The mean annual revenue for the Toledo District between 1914 and 1920 was a paltry $6,889, and expenses were $18,373 – producing an average annual deficit of $11,484. Although the forestry sector dominated the colonial economy, the colonial state generated little revenue from forestry exports and did almost nothing to regulate forestry practices. Two major colonial reports signaled the danger of the overcutting of mahogany and dependence on forestry exports. The crisis arrived with the Depression. Between 1927 and 1932, the value of forestry exports fell by over 90 percent, and to compound the crisis, on September 10, 1931 a major hurricane devastated Belize City and much of the mahogany-rich forests of northern Belize.[24] When export earnings collapsed, officials turned to London for Imperial grants. But at the request of the BEC and large estate-owners that owned most of the colony's wealth, the state reduced taxes on timber and chicle exports. To make up for this lost revenue, the state increased taxes on small landowners and peasants. London's concern that the colony would remain a major drain on resources led to a study of the colony's finances, conducted by Pim in 1932. He sought to impose strict limits on all state spending except for that related to agriculture and forestry in order to cut "the administrative organizations to the utmost extent," but without "postponing the prospects of development, both of agriculture and of the forests, which is essential if the Colony is not to remain a burden on the Imperial Exchequer."[25] To protect England from the burden of sustaining the colony's welfare, state spending must be reduced and exports must increase.

The limited data that are available from southern Belize give us a sense of the subsequent austerity and expropriation. In 1926–27, the District Board of Toledo brought in a revenue of $1,868, mainly from land and alcohol taxes.[26] After the price of Toledo's exports crashed, tax revenue more than *doubled* by 1931–32. The 1930s are one of the few decades since the establishment of a capitalist state in southern Belize when state revenue consistently surpassed expenditure. Depression, hurricane, and war accentuated weaknesses in the colonial accumulation project, which depended on the export of mahogany to generate foreign exchange and investment. Almost all of the land was owned by a few estate companies. Low land taxes gave no incentive to these companies to invest in developing their lands or secondary industry. Timber was exported essentially unprocessed; astonishingly for a "timber colony," the first sawmill in Belize was not built until 1932.[27]

Agriculture had always lagged far behind forestry as the major source of export revenue. Between 1882 and 1885 an average of only 12,661 acres of land were cultivated in the entire colony. Three main factors limited the development of agriculture: the lack of capital in the colony; the tiny internal market; and the land monopoly. Capital scarcity was both a cause and effect of underdevelopment in agriculture.

Under these circumstances, only the state could reorganize the means of production through agrarian reform, as occurred elsewhere in Latin America during the twentieth century. World War I restricted the colony's access to food imports, forcing the question of state agricultural and food policies. Studies by colonial agricultural scientists Sampson (in 1929), Smart (1929), and Stockdale (1932) all called for the state to create agricultural stations, but they focused on the need for the state to educate farmers. The colonial state thus interpreted the underdevelopment in agriculture as a result of a lack of knowledge by peasant farmers.

The landed elites who held the near monopoly on land and the logging companies and merchants that profited from the urban proletariat's food dependency (especially for imported flour) inhibited the development of a large agriculture sector and perpetuated the colony's import dependency. The land monopoly prevented middle and large peasant households from growing into commercial production for the internal market. A thoroughgoing land reform would have been needed to stimulate dynamic growth, but the colonial state avoided agrarian reform and land taxes that would have provoked the large timber companies and landowners.

We should pause to consider the long-term effects of the accumulation project of the colonial era on the development of productive capacity in the south. Belize's position in the colonial system was such that the potential profits from resource exploitation peaked during the mid-century mahogany boom, the colony's strategic importance was negligible, and the indigenous resistance to colonial intervention was considerable. The rates of capital accumulation associated with mahogany booms of the mid-nineteenth century and the 1920s never returned, and what value was accumulated flowed out to England and the US (British capital never substantively invested in the forces of production). All this contributed to a profound underdevelopment of the forces of production and state capacity. Although forestry exports gave way to agriculture in the 1950s behind expanding exports of citrus and sugar, the state never committed itself to a thorough agrarian reform, and the basic colonial structure of the economy has not changed.[28] Today the Toledo District continues to lag in agricultural production, and state capacity is the lowest in the country. Most Maya people still lack secure land tenure, and the area remains one of the poorest in Central America. The present-day demands for indigenous land rights – to which the state offers capitalism qua development as a salve – must be traced to this period of colonial extraction.

Hegemony, Settlement, and Territorialization

The ongoing struggles for indigenous land rights are strongly rooted in colonial political economy. These struggles largely concern lands that were claimed by the colonial state and institutions that were established to "settle" the Maya and win hegemony over them: in particular the alcaldes and Indian reservations.[29] These were created in the wake of the major mid-century conflicts with Mayas in the north and west of the colony, a period when the foremost concern of the colonial state was the territorialization of its space. Lieutenant Governor Longden argued for Indian reservations in 1868 in these terms:

> There are upon the Sibun River some villages inhabited by Indians, and until last year there were similar villages in the Western District, San Pedro, Santa Cruz, Chumbalche, San Jose, Nranjal, Quam Hill, etc., – several of these villages are situate upon the Lands claimed either by the British Honduras Company or Mssers. Young Toledo & Co., but wherever they are situate on Crown Lands I think the villages and a sufficient surrounding space should be reserved in the hands of the Crown for the use of the Indians, – no marketable titles being issued to them to enable them to dispose of such lands, – but the land being divided amongst them, from time to time, by the Alcalde or Chief man amongst them, as may be most convenient.[30]

This letter comprises the earliest attempt to justify a reservation policy. Three points are important to note. First, the question of rule by "Alcalde or Chief man" was tied to reservations from its first inception; these institutions were linked in colonial policy. Although alcaldes already existed in Maya communities, their precise powers and positions would change as a result of their incorporation into colonial "local rule" policies (see figure 1.4). The second point is that these policies were intended to address the problems faced by the two largest landholding and timber companies in the colony. Since the Mayas happened to be found by Europeans "upon the Lands claimed either by the British Honduras Company or Mssers. Young Toledo & Co.," the colonial state took up the responsibility of settling them somewhere. In light of the battles fought between the state and the Maya of the previous two decades, Longden desired settlement to reduce the likelihood of further Maya attacks. Third, Longden specifies two key provisos to the argument that "the villages and a sufficient surrounding space should be reserved . . . for the use of the Indians." On one hand, the land should not

Figure 1.4 Photo of an Alcalde's court, ca. 1948

Source: Annual Colonial Report for 1949. The photographer is not credited.

be held by the Mayas but must remain "in the hands of the Crown." On the other, the Mayas should have no means to convert the land into "marketable titles" that may "enable them to dispose of such lands."

In the spring of 1884, the government sent despatches to the colonial office on the subjects of the "appointment of Alcaldes in Indian Villages" and the "working of Crown Lands Department."[31] Although these three despatches were treated as independent concerns, they reflect a common challenge: to win hegemony over the Mayas. In September 1885, the Colonial Secretary, Henry Fowler, wrote to the Secretary of State for the Colonies: "our relations with the differing tribes of Indians on our frontiers are at present of a satisfactory character, and I see no reason to anticipate any change, provided the good understanding that has been established is encouraged, and some pains are taken *to cultivate the goodwill of the Indians*."[32]

Fowler's emphasis on cultivating goodwill marks an important shift from a mode of colonial hegemony that emphasized consent and territorialization more than military power. It was during Fowler's tenure as Colonial Secretary that the colonial state extended in southern Belize. When the state found unknown and ungoverned Mayas in the south, they moved to create state institutions that could win "the goodwill of the Indians." Colonial Secretary Fowler's despatch on the alcalde system to the Secretary of State for the Colonies in London offers a lapidary statement about the state's approach to the Maya; the challenge, as he put it, was to convert "the natives...from passive and indifferent subjects into loyal and willing subjects."[33] Fowler's letter goes on to detail the merits of the alcalde system: "the natives would appreciate a jurisdiction exercised over them according to their native customs. Leaving them to their own devices, or attempting to govern them directly by means of Magistrates and negro policemen has not worked satisfactorily." As evidence, Fowler cites a report from the District Commissioner (hereafter DC) of the Western District:

> The payment of salaries to the Alcaldes of the various villages causes great satisfaction and the anticipation of a staff of office and a flag to be hoisted before the Alcalde's house on Sundays and Fiesta days makes them feel that they will not be inferior in display to their neighbors in the Republics – Without these advantages it would be difficult to exact any service from the Alcaldes or to support their authority among *a people so childishly dependent upon ceremony.*[34]

Fowler argued that incorporating the alcaldes would bring several advantages. The alcaldes would be less expensive than police.[35] They could maintain order in rural areas by projecting the power of the colonial state through the manipulation of symbols. Thus, when Governor Harley wrote to the Colonial Office to express his thanks for their support of the recognition of the alcaldes, he proposed that "a staff of office – a cane – similar to those issued to friendly Chiefs on the Gold Coast, to be held during their tenure of Office."[36] As much as the alcaldehood was regarded as a "native" institution, in the sense that it already existed in Maya communities, it involved extending colonial power into rural Maya communities where the state could not go. By recognizing the alcaldes as indigenous leaders – as leaders who derived some authority from an indigenous political institution – the colonial state transformed them.[37] As with the establishment of reservations, the alcalde policy was intimately tied to the territorialization of the colony, fixing of boundaries, and delineation of Crown land.

By 1888, these policies had been accepted in London, but there were still no Indian reservations, though provisions were made for such in the Crown Lands Ordinances of 1872, 1877, and 1886.[38] Their materialization was delayed by two questions: whether the Mayas actually would pay land taxes on the reservation lands (the Colonial Office wanted to be sure that they did) and whether the lands for the reservations were actually within the colony.[39] To prepare the reservations, the Surveyor General wrote a report[40] and mapped the proposed reserves (see figure 1.5). Three Indian reservations were proposed: one in western Belize, near the Cayo, to be located "between the two branches of the Belize River... this would cause the various Indian communities scattered about the Western Frontier to settle within the reserve."[41] The second was to be in the north on Crown lands. The third was in the south, for San Antonio.[42] The map gives important clues about the spatial order imagined by the colonial state in 1888. The three proposed reservations not only touch the border in each case; they are well separated from each other on the periphery of the colony, which was centered on Belize City. In each case, the reservation is figured as a small, black rectangle – a *container* set aside for the Mayas – on the territory's margins (see figure 1.6). The reservations were thus imagined as spaces that made the colonial territorialization of Belize viable: spaces that would reduce conflict with the Mayas and naturalize the national borders by settling the Maya *inside* the territory of the colony.

Although the colonial state hoped that the Maya would stay within the borders of these three reservations, the state had no means (apart from the alcaldes) to actually police the borders. In southern Belize, colonial officials responded to the fact that Mayas refused to settle in one place by creating new reservations where they found Maya communities. Over the course of the subsequent two decades, the mismatch between the state's spatial order and Maya livelihoods led to the creation of several new reservations.[43] This practice continued up through the 1930s, until the Interdepartmental Committee on Maya Welfare signaled the end to creating reservations and a shift toward a new approach: to incorporate the Maya into the life of the colony and teach the Maya about their "connection with Britain and the Empire" in order to make them "Empire conscious."[44]

"The Indian will Require Patient and Sympathetic Treatment"

As the Depression set in and taxes were *reduced* for the large timber and chicle firms, the peasants of Toledo were asked to shoulder the burden. With one-fifth of the population, Toledo had the greatest number of households paying land taxes – mainly Maya and Garifuna peasant

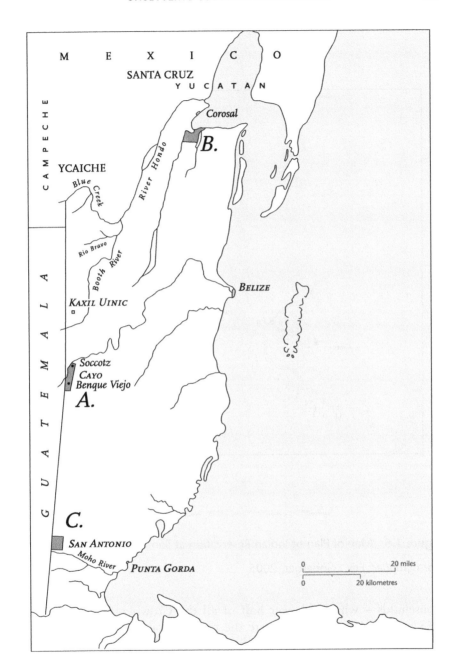

Figure 1.5 Map of Plan of Indian Reservations, 1888

Cartographer: Eric Leinberger, 2005

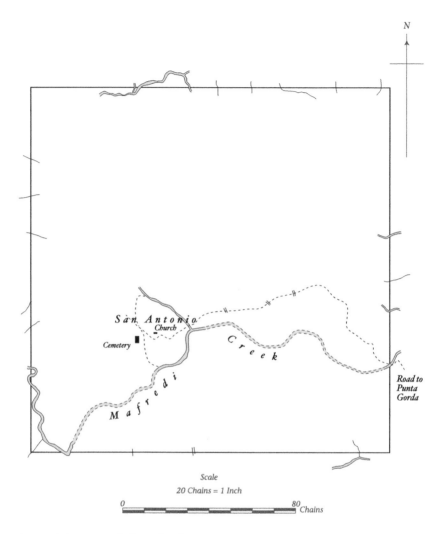

Figure 1.6 Map of Plan of Indian Reservation at San Antonio, ca. 1910

Cartographer: Eric Leinberger, 2005

households – who paid over half of all the Crown rents collected in 1932–1933. Although they were the poorest households (measured in cash income), the Maya of Toledo paid the highest *rate* of land taxes in the colony. In 1933, one official argued that land taxes were unbalanced: "A large proportion of the arrears are due from some of the most important land-holders in the Colony, and the criticism that they have been treated with greater consideration than the petty tenants of the Crown lands is therefore not without substance."[45] The Land and Property Tax Ordinance set the annual tax for Mayas cultivating Crown lands in Toledo at $10 per year, "out of all proportion to land rent elsewhere which was 30 cents an acre." This amounted to at least two months of labor at the prevailing wages – presuming one could find work.[46] After 1929, wage-work was hard to find, and prices for agricultural goods sold by Maya farmers crashed. Many peasants tried to avoid paying taxes, but the state vigorously pursued Crown rents, often threatening criminal suits. The rate of criminal cases per capita was 63 percent higher in the Toledo District than the national average, with a "large majority of the Criminal Cases tried summarily... for recovery of arrears of Crown rents."[47]

By March, 1932, the situation had grown so severe that more than a hundred Mayas walked from San Antonio to the Catholic Mission in Punta Gorda to plead for assistance. The Bishop wrote to the colonial government: "Some sixty Indians from San Antonio came to me about an hour ago – as a delegation of the whole pueblo, over one hundred being in town here right now – to explain their distress." He described the dilemma:

> The general depression and consequent want of a market for their usual produce – hogs and agricultural produce now esp. beans – makes it impossible for many to pay their rent to the Government ($10 a year). Many are in arrears for 1931. Instructions have been received by the DC from Belize... not to issue any permits for cultivation for 1932 to those who have not paid the arrears of 1931. Many have been summoned, the cases being heard right now whilst I am writing this. The Indians are *willing* and *anxious* to do all in their power to pay. To make the necessary money they want to ask for two things: (1) a market for their beans etc. be taken in kind. . . . The poor men came here backing their sacks the 25 miles & can't sell them after not for 5cts a quart. (2) Work from the Government on the San Antonio road. Both the San Antonio & (branching off) the 10 miles San Pedro Columbia road are in sorest need of work. They would gladly work for 25 cts a day and rations much cheaper than the work the Government does with its usual workers. – The money would go back to the Government in form of taxes.[48]

In response, the state asked the DC for a report on the status of the health of the Mayas: "The Bishop... tells me that the people in San Jose are starving.... It might be worth while taking payment in kind...."[49] When the Council asked the DC of Toledo to investigate "the economic condition of the Indians in the Toledo District,"[50] the DC duly reported that "the Indian has not yet felt and realized the full pinch of the world's depression." Since the Mayas "have invariably paid up their occupancy and habitation fees without a quibble," he reasoned, "the payment of habitation or occupancy fees [cannot] be waived." And since conditions were worse in Guatemala, the Maya had no choice but to pay the state: "I feel absolutely certain that enforcement of the payment of the fees will not drive them across the border to Guatemala."[51]

As the local representatives of the colonial state, the alcaldes were placed in a difficult position. Some tried to use their authority to persuade the state to assist them. In February, 1932, the alcalde of San Antonio wrote to Taylor:

> We the poor and industrious farmers of this village get our income and our living chiefly through or from the sale of our pigs and our commodities. From last June to the present time, we could not have demand for our market, and consequently, we are unable to meet our obligation in paying our rents in due time. This is an unusual circumstance on our part as Government may know. We have our pigs, beans and corn up here in abundance, and no buyers would come and buy in great quantity as before, and no place where to go and sell.[52]

Taylor's response was to turn to the new Agriculture Department to assist in "finding a suitable market for the commodities."[53] But the Commissioner of Forests, drawing from research recently conducted by the nascent Forest Department,[54] argued that the Maya farmers were to blame for their problems: "the difficulty [is] not so much lack of buyers but the *refusal of the Indians to reduce prices*."[55] If the state assisted Maya farmers with access to capital or markets for their goods, he worried:

> The Indians would probably take advantage of assistance given to raise the amount of the occupancy fees only and then cease cooperation. The Indian will require patient and sympathetic treatment before he will produce regularly for an established market.... [O]n no account should the payment of occupancy fees be waived.[56]

Having been called to order for asking the agriculture specialists to assist him, Taylor rushed to reassure his superiors that "the question of waiving the occupancy fees has not arisen. These are being collected and defaulting occupants have been proceeded against by law."[57]

At least 18 Maya farmers were imprisoned in September, 1932, for failing to pay their $10 land tax.[58] Reviewing the decision to jail the farmers, a state official who approved their sentence said: "Yes in this case, but no more imprisonment for no payment." His concern was not that their imprisonment was unjust. It was that "taxes must be worked off."[59]

Conclusion

Driven from their ancestral lands, the Q'eqchi' and Mopan Maya of southern Belize encountered a less violent, but nonetheless authoritarian, colonial state. In the forests and fields where they produced their livelihoods, their lands and labor were partly incorporated into the capitalist wage economy of the timber companies and agricultural estates. These companies took possession of most of the colony through an act of primitive accumulation involving a pair of deceits: first, that the lands taken for logging were not already held by the Maya; and second, that the logging claims should be converted into private property.[60] British colonialism meant the extraction of surplus through land occupancy fees, which forced Mayas to work at low wages for the timber, chicle, and agricultural companies that offered occasional employment. Unable to purchase the land that they paid for annually, the Maya communities saw no capital accumulation despite their years of labor and the production of a consistent agricultural surplus. Mayas resisted these forms of exploitation by avoiding taxes, moving frequently, living far from the state, and by trying to secure title to land through leasing.

The state complemented these processes of capital accumulation by attempting to settle the Maya. As a way of spatially fixing their subaltern economic position, the colonial state, logging companies, and the Catholic church collaborated to encourage the Mayas to settle in permanent communities where their resistance could be more effectively mitigated. The resulting hegemony that enabled colonialism was therefore doubly geographical. It was constituted on the basis of spatial forms of political power: settlement, land taxation, and territorialization. These in turn became hegemonic as they were naturalized and sedimented into common understandings of the geography of southern Belize. Notwithstanding the ongoing Maya resistance to colonial settlement (in its compound sense), the geographies of southern Belize came into being through colonial practices – at once political, economic, and geographical. As the ongoing struggles over these lands reveal, the very territorialization of southern Belize as such is an unfinished project. The colonial geographies of southern Belize remain to be unsettled.

Notes

1 Gramsci (1996): "History of the dominant class and history of the subaltern classes" (Notebook 3, note 14, p. 21). Gramsci revised this note in notebook 25, note 2.

2 Taylor, 1918, "Inducements to Indians to live in the Villages of the Toledo District." AB, MP 2427–18.

3 Under such leadership, subaltern (or "lower rank") social classes may give consent to their own exploitation in the absence of explicit repression. For Gramsci, the state is a site where hegemonic leadership is made possible and a locus of struggle for hegemony: the state "urges, incites, solicits, and 'punishes'" its subjects in order to create the conditions "in which a certain way of life is 'possible'" (Gramsci 1971: 247). Within this view, the capitalist state is a complex social relation that is both the product of particular struggles and contradictions of capitalist development, and also the most important arena for struggles to produce hegemonic projects (see Jessop 1982b: 100).

4 Texts written in the first decades after British Honduras became an official colony served to produce "the colony" and "British Honduras" as a space of resources. The history of this object thereby becomes told as an unfolding of knowledge about these resources. See especially Cockburn (1875); Fowler (1879); Morris (1883); Gibbs (1883); Bellamy (1889); Avery (1900); Lucas (1905); Collet (1909); Burdon (1927). There is a double move at work in the expression "the geography of Belize": this sign indicates both the mode of description of the surface of the space of "Belize" and the facts of the space so described. The joining of the discipline and its subject in a common sign contributes to the apparent naturalness of both the facts themselves and the proper process for deriving them. The challenge is not so much to deny the existence of geography but to read it always as a verb, as a process of becoming: not the world itself, but the production of worldliness. See also Greene (2005).

5 Lucas (1905: 320).

6 Spain opposed the presence of these settlers but had no interest in going to war with England to drive them out. The practice was legalized after the 1763 Treaty of Paris, where Spain recognized the rights of British settlers to cut and export logwood from Belize but maintained Spanish sovereignty; the Treaty of Versailles (1783) extended British control over Belize on the land between the Sibun and Hondo rivers. The Spanish claim to Belize effectively ended after their 1798 defeat at the Battle of St. George's Caye. Although the prohibition on agriculture did not prevent British settlements, it slowed the development of capitalist agriculture.

7 Shoman (1995: 193). The small rivers of southern Belize were a second factor inhibiting the colonization of the area; the absence of deep rivers

capable of navigation discouraged logging expeditions. The earliest British colonial state institutions date to 1765, when the Location laws were formulated by the settlers to formalize their rights to land at logging camps. In 1786 these rules were codified into Burnaby's Code, the first civil law. The first Superintendent of British Honduras, Colonel Despard, was appointed in 1786. The first constitution and Legislative Assembly were established in 1854. Belize was declared a colony in 1862 and became a Crown colony in 1871. Although British Honduras was governed by a colonial state before 1862, it was dominated by the settler community and not the Colonial Office in London. In this chapter, by "colonial state" I refer to the ensemble of institutions that ruled British Honduras after 1862. The first Magistrate for the Southern District (which comprised the area south of the Belize District) was appointed in 1865. The Southern District was divided into the Stann Creek and Toledo Districts in 1882. The best source on colonial British Honduras is Bolland (1977).

8   It is beyond the scope of this chapter to discuss the Manche Chol, whose presence in the area is documented in the Spanish records. Recognizing Chol loan words in the Mopan and Q'eqchi' spoken in southern Belize, Maya historian J. E. S. Thompson reviews the records on Spanish efforts to pacify "Chol country" in the 1670s (Thompson 1938). As he explains, in 1677 a Dominican father named Joseph Delgado was sent from the Bacalar to find an overland route north to Merida – a trip that brought him within present-day Toledo: "In 1677 Father Delgado and companion traveled down the Yaxal (Mojo [Moho]) River from Pusilha, and struck northwards along the coast of British Honduras." On June 7, 1677, while on the banks of a branch of the Moho River, Delgado wrote a memorandum describing the settlements and the land through which he passed (Thompson 1988: 35–9; see also Thompson 1938). His memorandum specifies the existence of a Maya community called *Cantelac* between San Antonio and Pueblo Viejo, near the "*Yaxal*" (i.e., *Yax Ha'*) or Moho River. Elsewhere I have argued that this memorandum contributes to the pattern of evidence that suggests that the present-day village of Santa Cruz is very near the probable location of a Maya village called *Cantelac*, probably home to Mopan-speaking (and possibly Manche Chol-speaking) people at the time of the arrival of the Spanish in the New World (Wainwright 2007). See also Jones (1998), Wilk (1997).

9   In 1859 the Superintendent of British Honduras said of the Mayas: "We know but little of these people" (Seymour 1859, cited in Burdon 1935: 221–2). Seymour became the first Lieutenant Governor of British Honduras in 1862. What they did "know" was that Mayas might attack logging companies or burn mahogany-rich forests. Speaking of the Mayas of northern Belize, Seymour says that they "have learnt to respect the mahogany tree in their clearing operations so there is peace between them and our woodsmen."

10 Ibid. This statement reaffirms that the area had not been explored, as it is not the case that the "line of water-shed" of the Maya mountains "forms the division between ourselves and Vera Paz." In many of the maps of this era, the divide of the Maya mountains is drawn along a north-south line. Also note that the verb "to contain" is crucial here, since it allows Seymour to suggest that the land is the *container* of "certain inhabitants" as opposed to their property or homeland (or simply a space that is already inhabited).

11 The early District Commissioners took no special interest in finding all the Mayas, let alone visiting the largest communities. The few reports written by colonial officials between the 1880s and 1920s suggest that trips outside of the three main coastal settlements (Monkey River Town, the Toledo Settlement, and Punta Gorda) were rare and brief. Reports by state officials as late as the 1930s indicate that trips to the rural Mayas villages were on the order of one brief tour per year to the main villages.

12 As the head of Alta Verapaz complained, with Mayas fleeing "it is very difficult to collect hands for agriculture" (Solórzano 1977: 288, cited by Handy 1984: 288).

13 Although most Mayas were living outside of recognized towns, the 1891 census counted 1,343 "people of Central American origin" in Toledo. At the time of the April 1891 census, rural southern Belize was an uncolonized region where timber companies were departing after over-cutting the forest. The 1921 census counted 2,169 Mayas. Many of the Q'eqchi's who migrated in the 1880s settled on land owned by German planters, who produced coffee, cacao, and nutmeg for markets in Belize City and Germany. But not all the Mayas settled on Cramer's lands, or in permanent communities. Maps from the area are marked with references to *milpas* and "Indian trails" along the rivers, suggesting that some Mayas refused to "settle" and sell their labor, preferring to make a livelihood off the land. See also Wilk (1997); Grandia (2006).

14 African slaves were brought to Belize via Jamaica by British settlers as early as 1720. Although the African slave trade was abolished by Parliament in 1807, slavery was not abolished in the colony until 1834, and the system of advanced-wage payment, combined with the few political or economic options for mobility among former slaves, insured that labor conditions changed only slightly after the end of slavery. Substantive changes in the work in the forestry sector did not occur before the decline of the forestry sector in the 1930s. Both slavery and wage slavery were consistently resisted; slave revolts are recorded in 1765, 1768, 1773, 1795, and 1820. On slavery in Belize, see Bolland (1997); also Shoman 1994: 43–63.

15 I found no records of exports from Toledo during the mid-nineteenth century in Punta Gorda, the Archives of Belize, or the Public Record Office. Such records are undoubtedly missing because there were no colonial state institutions in southern Belize at that time. Charles Wright once noted that there was a mahogany boom between 1825 and 1835, when land along the

rivers near the coast were first sold in large blocks, leading to the first major round of speculation (Wright, ACS. 1995. AB, CHW/20). This is plausible, but I have found no evidence to support Wright's claim.

16 The map shows the locations of the concessions: one in the southwest corner of the colony, next to the Sarstoon; four on the Moho River; five on the Rio Grande; one along the coast, just north of Punta Gorda; four along the coast to the north of the Rio Grande; and eight on the Golden Stream. In each case, the mouth of the river is reserved as Crown land, as are all of the lands to the west of the last concession, which on the Temax, Moho, and Rio Grande end about 10 miles from the coast. Note that the maps, like others from early in the century, are spare with details about the topography or geological features of the Maya mountains in the interior. This is further evidence of the lack of geographical knowledge of southern Belize by the colonial state. (For a case study on the production of geographical knowledge in southern Belize, see Wainwright and Ageton 2005.)

17 A series of maps in the archives from the 1880s reveals that this process entailed sending a government surveyor to the area to determine the boundaries around logging concessions granted in Belize City and later converted into private estates. See: Anon., 1887. "Plan of boundary of the Crown and B. Cramer & Co. on Moho River," AB, map, no catalog number; Anon., 1889. "A Plan for surveys of the Tidmash river made at the request of B. Cramer," AB, map, no catalog number.

18 Slavery was formally abolished in 1834, but as Bolland has documented, the social relations in the mahogany industry changed little with the end of slavery: see Bolland (1977, 1997).

19 Shoman (1995: 193).

20 Ibid., p. 197. In 1933, 6 percent of the landowners in Belize owned 97 percent of the land.

21 After Belize became a Crown colony, a Crown Lands Ordinance was passed that led to the reacquisition of some lands by the state. The law was most effective in southern Belize, where most Crown lands were accumulated. By 1900 over half the land in southern Belize was Crown land.

22 Based on maps from the 1880s and Minute Papers from the 1910s in the AB, it appears that Cramer owned almost every private parcel of land in southern Belize at one point, but held few of them for long. He speculated in land and tried to subdivide his parcels for sale or lease – often to peasants with Maya surnames.

23 Bananas were a major agricultural export from Belize between the 1890s and the early 1920s. Although the industry was centered around the Stann Creek valley during this period, the Toledo District was deeply involved in the banana trade. The industry was destroyed by Panama disease, first introduced in 1914. On the history of the banana industry in Belize, see Moberg (1996). Chicle exports from Belize to the US crashed with the Depression, with exports falling from over 4 million lbs. In 1930 to

1 million in 1932, and revenue from chicle duties crashing (from Bz $61,000 in 1930–31 to Bz $7,000 in 1932–33). Cacao exports also collapsed. Throughout the 1910s and 1920s cacao was grown in rural Maya communities (25,000 lbs of cacao were exported in 1913), but the price of cacao crashed during the Depression, falling from 50 cents to 4 cents a pound between 1928 and 1930. Plans to export cohune oil dried up in the same period. In 1929 the Tropical Oil Products Company purchased a 46,000 acre estate and attempted to export cohune oil. The project created a settlement of several hundred workers in rural Toledo. It had just become operational when cooking oil prices crashed, and the scheme was abandoned around 1939.

24 Grant (1976: 62).

25 Ibid.

26 The District Board was comprised of nine people – mainly elites appointed by the Governor, and headed by the DC.

27 Government of Belize, 1934, cited by Camille (1994: 122). The mill was built by the BEC in Belize City for milling logs into boards.

28 Before the 1950s, the state's lack of interest in promoting commercial agriculture was usually explained with reference to its peculiar colonial history of dependence on forestry. Successive PUP governments have made some progress in reducing Belize's food import ratio: food imports accounted for 31 percent of imports as late as 1974 and only 14 percent in 2000. This decline masks the fact that most of the processed food consumed is still imported and much of what is eaten – especially chicken, wheat, and soft drinks, "Belizean" staples today – is typically imported and/ or controlled by expatriate or Creole families. On Belize's development strategy 1960–1980, see Palacio (1996); on the colonial roots of Belizean cooking, see Wilk (2006).

29 On the history of the Indian reservations, see Bolland (1987); Berkey (1994).

30 Governor Longden, 1868. Despatch No. 39 of 1868, cited in Despatch No. 8 of 1884, PRO CO 123/172.

31 PRO CO 123/172. The alcaldes despatch was written on January 24, 1884, the same day as a despatch announcing new measures for securing "prison discipline." The control of the Mayas and the prisoners were debated at the same time.

32 Fowler, H. 1885. PRO CO 123/1885/Despatch 134, letters discussing the "grand revolution" of Yucatan. Italics mine.

33 Fowler, H. 1884. "Appointment of Alcaldes in Indian Villages," PRO CO 123/172.

34 Millson, A. 1883. "Report on the Western District, 7 December 1883", PRO CO 123/171, p. 7. Millson elaborates:

> This probable influx of Indians [from the Guatemalan side of the border] makes it advisable that the question of an Indian Reserve should be considered. It would be unadvisable, and to some extent dangerous to eject those at present

settled in the towns above mentioned, but, at the same time to avoid future difficulties from the encroachments of cattle from the mahogany works upon their Milpas, as well as to provide a refuge for those who are desirous of settling in the Colony, it would be wise to set apart some well defined tract of Government Land for their undisturbed possession; such lands might either be chosen, in the neighborhood of San Antonio, stretching back into the Great Southern Pine Ridge, or between the two branches of the Belize River from The Cayo to Garbutt's Falls, and more northerly position being in too close a proximity to Yaloch and Ycaiché. The chief objections to selecting the land between the River branches for such a Reserve are its proximity to The Cayo and other Creole Settlements, and the somewhat limited extent of land available for *people of such nomadic habits as the Indians*, who, *depending as they solely do upon the practice of their Milpas*, and totally unacquainted with any means of fertilizing the ground they cultivate, are *forced to move from place to place* as the soil of their plantations becomes exhausted. (My italics)

From the earliest stages on planning the alcalde/reservation system, the colonial state hoped to transform Maya agricultural practices. Colonial rule and the formation of hegemony called for such development. I take up the discourse on Maya farming in the next chapter.

35 In his estimates for 1923–24, the Toledo DC noted the costs for the District staff. More was spent on the DC's stationary ($25) than a second alcalde would have received for an annual wage (McCall, T. 1922. "Estimates for the year 1923–24." 14 November 1922. MC 956 (this MP is misfiled: it is actually 3051–22).

36 Anon. 1882. "Indian Affairs". PRO CO 123/168, 30 December 1882.

37 Alcaldehood thus became a site of the intense and polyvalent reworking of colonial power, as it remains today. Today the Toledo Alcaldes Association is a registered NGO.

38 Fowler, H. 1884. "Crown Lands Department. Forwards report on working of by Surveyor General." PRO CO 123/172, Despatch 40/1884. Fowler elaborates:

> The Committee of the old House of Assembly recommended certain lands to be secured to Indians and Caribs and sufficient lands be reserved for Churches, cemeteries, and other buildings. The Crown Lands Ordinance of 1872 provided for "lands being reserved for the use of Indians and Caribs and for permit being issued renewable yearly to occupy lots," and Section 24 of the Ordinance provides for "Indian and Carib villages being surveyed and reserved for the use and enjoyment of Indians and Caribs". . . .
>
> I am very strongly in favor of forming Carib and Indian reserves as a mere act of justice on the grounds of former recognition of the claim of these natives. As regards the Indians the only step taken in the matter was in 1880 to lay out two Indian villages and to sell the lots at $2.00 each. This was approved by Sir M. Hicks-Beach in despatch No 26 of 7th April 1880 but nothing further has been done. The whole question requires careful consideration and in a systematic manner. I respectfully submit that due

provision should be made therefore by a simple amendment authorizing the Lieut-governor in Council to reserve portions of Crown Lands for the use and enjoyment of Indians and Caribs subject to the sanction of the Secretary of State and to make regulations for the management and government of such reserves subject to the approval of the Secretary of State.

At this point in the text he elaborates on the 1877 Ordinance:

In 1877 an Ordinance to amend the law in this respect was passed on the grounds that the Caribs had disregarded the above provisions and they were repealed. Nothing is said about the Indians, and the Lieut-Governor was authorized to lease lands within Carib reserves for any term not exceeding 10 years at such rent as he may deem proper, or grant the same in fee for not less than 5/- per acre. This Ordinance was repealed by the present land ordinance in which no mention is made of either Carib or Indian reserves, so that these natives are now subject to the same provisions as other people. Such rights as they may have secured under the previous laws, although stipulated were not to be affected by the present law, in the absence of any records thereof, have been simply ignored. The imposition of a rental of $2.00 for each Carib house in Stann Creek in 1878 nearly caused serious disturbances. The District Magistrate reporting that the Caribs threaten to make "a day of it" if the Crown bailiff attempted to "enforce the rents," and Mr. Downer "thought it high time to cease moralizing with such savages – that it is common talk among the Caribs that they do not know why they are to pay rent as Stann Creek is their place, that long ago the land was given to them and they settled the place." These rents have been collected as best they could since, but there is little legality for such proceedings since the repeal of the 1877 Ordinance by the 1879 Act.

39   Consider the following notes from a November 1888 minute prepared in London:

[T]here are in the Northern District Indians who consider it belonging to them, land which when the Mexican boundary is settled will be part of the Colony. If so, it may be necessary or expedient to provide a reserve for them either at the place marked B on the map annexed to the written report or somewhere else in the North West of the Colony when the Boundary Convention with Mexico is signed. . . . With regard to the proposed reserve at the place marked B in the map I see no reason why Indians already renting land from private persons should have free grants there. It may be convenient to settle there any Indian that may come into the Colony from Yucatan, but I am not prepared at present to approve of their being given the land or allowed to occupy it rent free. I should be glad to know whether the ordinary tax on cultivated land is paid by the Indians on their milpas. It should be ascertained whether San Antonio is within the Colony and if it is within the Colony a reserve should be marked out.

Although it was later determined that San Antonio was within Belize, the first map of the Indian reservations shows the community to be on the border with Guatemala.

40 Miller, W. 1888. "Proprietary Rights of Indians." PRO CO 123/190, Despatch 129, 28 September 1888. Miller's report is enclosed in a letter sent by Hubert Jerningham to Lord Knutsford on September 28, 1888.

41 Ibid.

42 See Miller, W. 1888. "Report on Indian Reservations." PRO CO 123/190. The state had little idea in the 1880s where Maya communities in southern Belize were located and whether they were even "within" the territory of the colony. Miller writes of San Antonio: "[T]hey pay nothing for their lands. There is some doubt in my mind as to whether San Antonio is actually within the limits of the Colony as the journey to it occupies three days from Punta Gorda. The road is said to run due West and the Colony is only 30 miles broad at that point."

43 ICMW (1941). Some Mayas also leased land from the colonial government. Leasing appears to have been common since the 1890s. The process of applying for a parcel of leased land required the supplication of the would-be leaser to the DC, who would then submit the application to the Surveyor General. Maps from the archives indicate the leasing occurred in the late nineteenth century, and Minute Papers from the early 1900s indicate the same. For instance, in 1905 Victoriano Pow applied to lease 100 acres of Crown land to the north of the San Antonio reservation. When the Surveyor General checked his files, he found that three of the acres Pow was applying to lease were already leased to another Maya farmer. The Surveyor remapped the proposal and granted 97 acres to Pow. This and other applications by Mayas from this period were treated as routine. It is difficult to estimate the extent of leasing, but in 1921 there were no fewer than 120 applications to lease land in the Toledo District, and 24 of these came from people categorized racially as "Indian." (Taylor, J. 1922. "Annual report for the Toledo District for the year 1921." AB, MP 1766–22. Taylor notes: "as soon as they get here [from Guatemala], and find out about land matters, they apply to lease.") The fact that Mayas could lease land underscores the fact that the state was mainly interested in settlement and the collection of land occupancy fees.

44 Ibid., Appendix I. Not incidentally, the ICMW report also called for a development policy aimed at replacing "the *milpa* system," an object I discuss in the next chapter.

45 Pim (1933: 43).

46 In his report on the District for 1921, Taylor reported that while calculating the acreage leased by Mayas since 1914 "would take up too much time … the increase is fairly large. Roughly, taking the Toledo District as a whole, and including the Monkey River – (which has no Indian population) – the value of the Country lands rent roll in 1914 was about $1,800, and now it is very close on $4,000." In 1921, when the rent roll for Toledo was $4,000, the population of the Toledo District was estimated at 5,242, of which at least 2,169 were Mayas. The latter would have comprised the majority of

land leasers. Taylor notes that the average monthly wage in 1921 for mahogany work is $15, and for agriculture, $10 (Taylor, J. 1922. "Annual Report for the Toledo District for the year 1921." AB, MP 1766–22).

47   Pim (1933: n.p.).

48   AB, MP 1060–32. We should be skeptical to take the Bishop's word on this (the church could not adequately represent the desires of the Maya). It is difficult to know whether the Bishop's claim that the Maya were *"willing and anxious...to pay"* shows that many Mayas consented to paying their land taxes, even under the most difficult of circumstances.

49   Anon. 1932. "Authority to write off amount for which Indians served imprisonment for non-payment of occupancy fees during 1931." AB, MP 2068–32. The problem was simply that "The Indians must learn that prices of primary commodities have fallen and that they can no longer hold out for fancy prices" (Anon. 1932. AB, MP 2068–32).

50   AB, Minutes of the Executive Council, May 11, 1932.

51   Alcoser, A. 1932. AB, MP 2068–32.

52   Anon. 1932. AB, MP 508–32.

53   Taylor, J. 1932. AB, MP 508–32.

54   Fearful that Maya peasants would cheat on their taxes, the Forest Department studied their production and consumption habits in 1936. The August 1936 research conducted by Forest Ranger Hope (compiled with commentary in Minute Paper 266–33) comprises the first thorough, modern census of the Maya of southern Belize. State sponsorship of research on the Mayas was from the outset oriented toward statistics and tax accounting. From this research we know that Maya households may have suffered serious losses during this period. The 1948 Annual Report for the colony reports the "abnormally high" death rates of 1931–1946 among Maya communities in southern Belize, which may well be attributed to the acute poverty of this period.

55   Ibid., my italics.

56   Anon. 1932. "Difficulties being experienced by the San Antonio Indians in marketing their produce." AB, MP 508–32.

57   Ibid.

58   The $180 they owed to the state was written off, as they had paid by their time in jail. (Anon. 1932. "Imprisonment of Indians for non-payment of occupancy fees for the year 1931." AB, MP 2068–32.)

59   Ibid.

60   Although the state is the largest landowner in Belize, private ownership of the land is highly concentrated. In 1977, 85 percent of the private land in Belize was owned by only 42 people (Bolland and Shoman 1977: 7; data from Lands Department land tax rolls).

2

The Matter of the Maya Farm System

This history will not be one of discoveries and of errors, it will not be one of influences and originalities, but of the history of conditions that make possible the appearance, the functioning and the transformation of . . . discourse.

Michel Foucault (1996: 66)[1]

In 1999 a new agricultural development project was launched in southern Belize, the "Community-initiated Agriculture and Resource Development project," or CARD. Like several earlier projects, CARD's aim was to reform Maya agricultural practices and accelerate development. The project is governed by a pair of loan agreements that were signed between the donors – the International Fund for Agricultural Development (IFAD) and the Caribbean Development Bank (CDB) – with the Government of Belize (GOB), where one finds the usual elements of a development loan package: budgets, plans, timelines, a statement about "gender mainstreaming" – the boilerplate of development writing. Notwithstanding their formulaic structure and unimaginative writing, these are texts with clear material effects, therefore worth close reading. Indeed they must be read because they are formulaic in ways that produce the authority of capitalism qua development.

At the end of the loan agreement between the Government and IFAD, we find an Appendix that may seem out of place. It is an anthropological statement, entitled "The *milpa* system and ethnicity and culture." It begins:

The *milpa* system. Most of the project target group base their livelihood on the *milpa* system. The Mayans practice the *milpa* system that is integral to a deep-rooted set of cultural patterns and beliefs. It consists of a shifting slash-and-burn system of cultivation. Its principal output is corn and beans as crops for home consumption, and *milpa* rice and cacao as cash crops.[2]

Not long ago, it may have been hard to imagine a loan agreement from a major development bank including an essentially ethnographic description of the "target group." Yet today it is not uncommon to find such statements woven into development project loan documents (albeit often, as here, in the appendices, literally as subtext). Concluding the loan agreement with such an appendix gives the impression that CARD's plans for social and economic change are grounded in a careful assessment of culture. For many, the inclusion of such anthropological sensibilities reflects a progressive shift within development thinking, since it suggests that local cultural practices will inform the work of this development project. But before we adopt this reading, or celebrate the inclusion of such an appendix, I suggest that we should ask: why are these statements ("Mayans practice the *milpa* system...") in the loan agreement at all? What is the history of this particular set of statements? And what effects do such statements produce?

This chapter begins to answer these questions by reading in the archaeology of the discourse about Maya agriculture. I track a key concept that structures the passage from the appendix: the Maya *milpa* system. We note that the word "system" appears here four times (and more elsewhere in this text); it is said to describe an "integral" and "deep-rooted set of cultural patterns." We are told that the Maya are very systematic people. Their system works to convert inputs – the forest that is burned, "cultural patterns," beliefs, and labor – to produce agricultural outputs (corn, beans, pigs, rice, cacao). These outputs are essential to the reproduction of the Maya culture and labor, and so the elegant system is complete – except for the matter of the forest, which, we are told, suffers from this "shifting slash-and-burn system."[3]

These words are not used here in an original or arbitrary way, but as part of a longstanding discourse about Maya agriculture. While the appendix presents us with a novel articulation of development planning with Mayanist description in a financial contract, the *possibility* of such an articulation of statements and development finance is grounded in a discursive formation that is roughly a century old.

Archaeology, Mayanism, and Agriculture

It is the emergence of this discursive formation – and its object, the "Maya farm system" – that will be my concern in this chapter. I explore the conditions that made possible its appearance in the 1920s by tracking a set of interlinked statements affiliated with three disciplines that, while apparently distinct, share a certain way of framing concepts and explaining practices: agronomy, archaeology, and anthropology. By so doing I aim to call into question the ontological status of the *milpa* as a Maya farm system. My premise is that this way of speaking about the Mayas was neither inevitable, based on objective truths about the world, nor mere fiction, as if invented by creative minds. Rather, the *concept* of a "Maya farm system" (hereafter MFS) has its own history. This history can be traced with the modest aim of trying to understand what exactly is spoken of, how this farm system comes to be essentially linked with what is "Maya," and how it emerges as a problem that requires the attention of the state. To make this argument I read texts from Guatemala and Belize, and draw upon the approach to discursive formations developed by Foucault in the 1960s. As he explained in *The archaeology of knowledge*, this entails studying "discourses as practices specified in the element of the archive,"[4] in order to map a distribution of statements that comprise a discursive formation. He called this system of possible statements the "archive"; thus the study of one of these discursive systems constitutes an "archaeology."

Foucault's approach was a critique of the idea that the historian could unearth the truth by digging below the surface of speech to the real ground of meaning. Foucault opposed the thought of defining the *arché-* as such (his use of the term "archaeology" was intended to be ironic) since it reduced discursive practices to the level of superficiality and resulted in conventional histories that produced hierarchical, multi-layered structures and measured facts by their relative attachment to the real:

> From the political mobility at the surface down to the slow movements of "material civilization," ever more levels of analysis have been established [by historians]. Beneath the rapidly changing history of governments, wars, and famines, there emerge other, *apparently unmoving histories*: the history of sea routes, the history of corn or of gold-mining, the history of drought and of irrigation, the history of crop rotation, the history of the balance achieved by the human species between hunger and abundance.[5]

The eye of the historian, Foucault argues, is attracted to the rapid and intense rhythm of change at the surface of human experience – at the level of government, famine, and war. These rhythms are then typically analyzed through some sort of structural analysis (even when it is not made explicit) of the broader rhythms of changing geographical patterns and "material civilization," such as sea routes and corn, that unveils the true roots of merely superficial historical affairs. With this sketch of the multi-rhythmic reading of history, Foucault announces that the archaeological approach rejects the view that there is an "apparently unmoving history" of material-geographical phenomena deep down below the surface of meaning (in the sense of everyday speech). There is no timeless, settled core underlying the surface of things. What this suggests is that even the "apparently unmoving histories" can be investigated and written.

When he wrote this passage on "apparently unmoving histories," Foucault may not have known that the history of corn and civilization had in fact been written – as, indeed, *the* story of the Maya. The history of Maya civilization and agriculture has been written in an empiricist mode, precisely as the history of "the slow [rise and decline] of 'material civilization,'" determined by the development of "corn, . . . crop rotation, [and] the balance achieved by the human species between hunger and abundance." I offer the following reading of the Maya farm system through and against such histories. I do so without proposing an alternative narrative and without promising that this reading will make visible any hidden subaltern histories. If this reading is justified, it is because the Maya farm system is fundamental to colonial discourse, and therefore crucial to interpreting the relationship between colonialism, development, and Mayanism in Belize. For capitalism qua development in Belize, the so-called "Maya problem" has long centered around the task of settling the Maya farm system.

Discourses of Agriculture Before the Maya Farm System

We begin by considering ways of speaking about Maya agriculture before the appearance of the Maya farm system as a discursive object.[6] In so doing we find that certain statements about Maya farming are iterated in ways that form definite patterns.[7] We can define two such patterns in the discourses about agriculture and the Maya before the 1920s. In accounts of agriculture from before this period, the productivity of agriculture is not attributed to systematic or essentially Maya practices. It is rather *in nature itself* that the abundance of life is found.

The first element of discourses on Maya farming is that their agriculture brings forth nature's spontaneous bounty. Where agriculture is analyzed and written about in relation to Maya people by early Spanish missionaries, their gaze it not focused on defining cultural traits or natural resources *per se*. Most descriptions speak of the Maya as "Indians," in the way that Spanish colonists would have referred to Nahuatl-speaking people in what is now Central Mexico. The space of their description is a landscape – a space of novel "Indians," plants, and animals. For instance, in 1574, Padre Gallego y Cadena wrote a text on the area now known as the Baja Verapaz of Guatemala; it describes the climate, rivers, mountains, forests, and so forth. The concern is not to define potential resources, however, but to produce something like a descriptive catalog of the space of nature. And so agriculture is spoken of always in relation with other signs of nature, such as climate and the verdant landscape:

> The weather is, for the most part hot, and the rest of the time almost mild where the village and most or almost all of the people are. Their fields, the forests, and the valleys are always green and full of flowers and whereas in Spain we see flowers in the month of May here they grow without exhausting themselves for the whole year.[8]

The central trope of Gallego's writings on Guatemala is the *abundance* of tropical nature.[9] His description includes lists of fruits and trees, differentiated by those that are and are not found in Spain. There are no specific descriptions of agricultural practices; only in a few passages does Gallego mention food production.[10] This element in also dominant in an 1841 text by Alonso de Escobar, a Catholic priest who traveled through the Verapaz region and wrote an "Account of the province of Vera Paz in Guatemala." His account of the region, subsequently translated and published by the *Journal of the Geographical Society*, describes agriculture in the region. He stresses the prodigious bounty of nature:

> On the north-west are the mountains of Chisec, anciently inhabited by the Indians now established in the Alcalá division of Cobán. In the same mountains the Indians of Cobán still grow their cotton and keep their plantations of achiote and cacao; *not that they plant or do much more than take advantage of the earth's spontaneous production.*[11]

We should pause here to compare this last line with the statements by Gallego. Escobar says that the "Indians" do not "plant" corn; they simply "keep" their plantations and enjoy "the earth's spontaneous production." The parallel to Gallego's expression, "'watch over' their *milpas*" is striking.[12] In such statements about the land and the Maya, there is no space for analyzing the connection between the Maya and nature, which are not, at this point, firmly differentiated phenomena. The distinction between the labor of the Maya and the environment in which they live is left ambiguous, if drawn at all.[13]

One reason for this ambiguity, from the perspective of the contemporary social scientist, is that the notion of systematic cultural practices is not dominant. Miranda, a Spanish priest who passed through Guatemala in the mid-sixteenth century, speaks of Maya planting practices but not in a way that suggests the work of a culture.[14] Miranda writes of "Indians" as his source of information about the natural wealth of the land and does not suggest that the wealth of the land is a product of Maya labor or knowledge. This is not to say that Miranda and other Spanish friars did not regard agricultural production as hard work – they often mention the labor invested by the "Indians." But their statements suggest a naturalness of their way of life as part of a naturally productive environment:

> Plant twice a year, once in April and harvest in October; this is their major planting season and in the forest that they cut down and burn, and this which they have less than twenty days in the summer, as they call it, when to burn, because once there is no sun [when the rains begin], they cannot burn, and if their seed is not planted, bringing rats to destroy it. This felling of new bush is done every two years, because it [the soil] cannot produce corn more than twice, and more than 10 or 15 years pass before they return to plant, and for this reason planting season is very labor intensive for them because during it they have to cut down many trees, and a lot of them are thick.[15]

When the Spanish write of the land in admiration of its natural bounty, and the Maya enter their descriptions, it is as "Indian" informants of the land. Thus, Artiga writes in 1699: "I have seen the corn harvest, and the Indians say that when they make a new clearing they build a house in it, and the milpa lasts for twenty years without ceasing to give two harvests per year."[16] The identification of "the Indians" here as inhabitants of the land and informants speaks to the productivity of nature ("without ceasing to give two harvests per year"). It presages an ethnological mode of writing yet to be consolidated.

We have seen that the early Spanish discourse does not register a strict separation between nature, labor, "Indian" inhabitants, and agricultural productivity. These elements are combined in a way that presupposes the productivity of nature and locates the "Indian" within a space of spontaneous, bountiful existence. Why should it be so? Were these ideas necessary in order to justify colonization? One could, perhaps, argue as much: by showing that the land-and-Indians were naturally productive, these descriptions could be read as texts describing the beauty and bounty of nature in the Americas, as prospects for colonization. But such a reading would reduce the play of discourse to a question of ideology. These texts were not produced to meet the requirements of capital or the colonial state, and the discourse of Maya farming is more complicated than this, more fluid – in fact, it radically changes at around the time that the British came to equate colonization with development in Belize. The relationship between discourse and colonization, in other words, is not instrumentalist.[17]

"Somewhere in the Depths of the Forest"

Before the twentieth century, the agriculture of the Mayas is spoken of in relation to a space that is natural, but where this naturalness is divided. It is not natural only in the sense of spontaneous abundance, but also in the negative sense – as an uncivilized, disorderly, unsettled space. The Maya are figured as a scattered people.

Let us return to Miranda. At several places in his text, Miranda describes the spatial organization of the fields and houses he encounters. For instance: "The rest have dilapidated buildings without harmony or order of wide streets but rather with some paths, like those made by deer. [T]he houses are fenced and closed to the sun and air."[18] This statement should be read against another, written almost three hundred years later, in a history of the Protestant Christian church in British Honduras and Guatemala.[19] In his 1850 text *The Gospel in Central America*, Crowe describes "Indian" livelihoods and settlements in terms similar to Miranda:

> The townships of the Indians are mostly scattered and irregular. They frequently occupy a whole valley or plain, or lie straggling for miles along the roadside, concealed from view among bushes and fruit trees.... The "Rancho" [a Spanish term for a Maya farm] is a mere temporary shed thatched with palm or other leaves, and is used as an alkilo by Ladinos and Indians.... Their cattle generally stray at large in the forests and pine ridges, or even in the roads and streets, and on this account the plantations of the Indians are mostly at some distance from their houses.... When not engaged in cultivating the soil, which occupies but a small proportion of

their time, the Indians make frequent journeys, carrying their wares and fruits to distant markets.[20]

For Crowe, the spacing of social life by the "Indians" appears disorganized and confusing. Their "scattered" settlements are *"irregular"* and "lie *straggling"*; in Crowe's eyes, they are disorderly and irrational. (And "the Indians" do not even stay there, but leave on "frequent journeys"!)

Similar statements are scattered throughout other texts of this period. In 1875, Cockburn wrote what is arguably Belize's first resource assessment. He surveyed much of British Honduras, which had recently become a formal British colony. Cockburn's account offers a systematic description of the ways wealth was produced in the colony, and so takes special consideration of agriculture. Despite his aims and mode of writing, however, he makes no mention of a Maya farm system when he describes Maya *milpas* and maize production. His only direct comment about Maya agriculture comes in a description of the town of Corosal:[21]

> The place is dotted over with several little plantations called "Ranchos" and "Milpas," 10, 20, to 100 acres in extent, where, besides the sugar cane plantations, corn, rice, and other provisions are grown, and sugar and rum manufactured in a primitive way.... They carry on their operations at a comparatively low expenditure, as their laborers are chiefly their own countrymen (native Indians) who are content with but little pay and no rations.[22]

Agriculture is still spoken of as a "primitive way," part of a landscape "dotted over" with plantations. We are still within the purview of Gallego, where the distinction between the landscape, labor, and agriculture is not firm, but rather an ensemble. The farm as a system does not exist. Indeed, in one striking passage, Cockburn gives a premonition of how colonial discourse will later encounter this ensemble. Describing sugar cane agriculture in northern Belize, Cockburn reasons that the soils there must be excellent because sugar cane grows well "in the face of the Rancheros' system of cultivation, or rather *want of system."*[23]

In the texts by Crowe, Cockburn, and Tozzer, a new element of this discourse is clearly present: the consolidating identification of a relationship between Maya *culture* and *agriculture*. It is through the specification of the particular spatiality of *milpa* settlement that we find the earliest specification of "the Maya" as a *race* – as well as a potential source of labor. Thus, Cockburn specifies that *milpa* labor is performed by "(native Indians) who are content with but little pay and no rations."

The process of discursively constituting the category "Maya" as a race is a slow, iterative one that is not (and cannot be) completed. But already in the nineteenth century, discourses about agriculture in Guatemala and Belize become increasingly racialized. Spaces of production are separated on the basis of groups that are treated, even sometimes defined, racially.

Statements about the Maya and their mode of living in the nineteenth century articulated British colonial views regarding racial categories as well. In June 1859, the Superintendent of British Honduras sent a report from Belize City to London concerning the Blue Book for the Settlement (which was not yet a Crown colony) for 1858. As part of that report he offered his observations about the races of the colony; replete with anxious speculations, the text proposes appropriate labor categories for each race.[24] Consider the following note about the Maya, called the "aborigines" of Belize:

> The aborigines deserve the priority of notice.... In the more civilized districts they are much mixed up with persons of other races, either as patient and silent domestic servants, or as useful members of the mahogany gang. More robust than the Spaniard, less addicted to pleasure than the negro[,] they are admirably adapted to the monotonous drudgery of logwood cutting which has principally passed into their hands. There are... other tribes of Indians within our borders who come in contact with civilization but once a year. They cultivate maize somewhere in the depths of the forest and fatten pigs whose surplus produce they annually bring to some village market, procure what they require, principally salt[,] and disappear again.[25]

Several strands of colonial discourse are interwoven here. First, "the aborigines" are treated as a distinct race. The basis of this distinction, and what gives them "priority of notice," is that they are not Spaniard or "negro," but rather the "descendants of the men who built the temples"[26] (we will return to this theme in the next chapter). Second, the "mixing" of the "Indians" with other groups is regarded with some ambivalence. Through the mixing they become laborers – "patient and silent domestic servants, or... members of a mahogany gang." Without mixing they lose out on the "civilizing" effects of living in cities. Further, such workers are "adapted to the monotonous drudgery of logwood cutting." The last and most important strand of colonial discourse here concerns civilization and space: the Mayas are those who "cultivate maize somewhere in the depths of the forest"; they are not settled, where they could "come in contact with civilization," but scattered.

This structure of statements about the Maya and agriculture can be read in popular accounts from this period as well. In 1883, Gibbs wrote one of the earliest general histories of the colony. He speaks of the Maya in essentialist terms, but not in terms of a system. When he addresses the question of "the Maya Indians" and agriculture – which he does only once in the text – he does so in order to note the alignment between the "inoffensive" racial essence of the Maya (defined in terms of "disposition"), their agricultural livelihood, and the ungoverned spatial distribution in the colony: "In disposition they are docile and timid and inoffensive.... They live industriously and inoffensively in villages scattered over the district, cultivating their patches of maize and pulse, their pigs and poultry."[27] We read that the Mayas are "scattered over the district," which in this text is not treated as a problem *per se*. While his remarks are imbued with a strong ethnographic paternalism ("they are docile and timid and inoffensive"), Gibbs does not say that the Mayas are missing out on civilization. Nor does he suggest that they live within something like a "Maya farm system."

We find the earliest *scientific* accounts of the resources available for the development of the colony of British Honduras in the colonial records of the late nineteenth century. Initially, however, these accounts did not break radically with the Spanish tropes about the "Maya" and agriculture. When, in 1883, Morris was sent to conduct a scientific survey of the agriculture of British Honduras for the colonial government, he aligns the geography of agriculture with racial categories, but does not speak of the Maya as having a particular farming system.[28] Agriculture and Mayanism, although affiliated, are not systematically co-constituted.

This pattern of speaking about the Maya and agriculture existed until the early twentieth century. By the end of the 1920s this discourse would be replaced by another. The tropes of spontaneity, naturalness, and productivity would be reorganized, new elements introduced, and new objects produced. In short, the discourse about farming and the Maya would become systematized by three empiricist, scientific disciplines: agronomy, anthropology, and archaeology. A new way of speaking about the Maya would appear, involving new concepts – system, culture, carrying capacity – arranged in a way that constituted a new discourse.

The Discourse on the Maya Farm System

In 1924, British Honduras was represented at the British Empire Exhibition in London, held to celebrate the Empire in the post-World War I days of growing anticolonial sentiments. A series of three pamphlets was prepared for distribution at the Exhibition that described British Honduras (while outlining prospects for investment). The first pamphlet provides a general discussion of the history and geography of British Honduras.[29] A second pamphlet describes the "Woods and other forest products of British Honduras." The third is entitled, "British Honduras: its agriculture and minor industries." In this third text we find a statement about the production of "Indian corn":[30]

> Indian corn is a very important food crop in the Colony for both people and animals. *The crop is grown on the "milpa"* system*, especially in the Orange Walk, Cayo, and Toledo Districts. The Indians are the principal corn growers, but the mahogany companies also grow large areas for the purpose of stock feed. The yields obtained are very satisfactory, the size of the ears is remarkably large, considering the primitive methods of production.[31]

This text carries a radically different tone compared with nineteenth-century discourses. Not only is corn production defined as a system, but the author, Aspinall, adds technical details about the local methods of production. He does so with an eye towards the limits of this system:

> The Indians select their seed for planting, but they do not select it in the field. Proper selection of corn would increase the yields considerably. The writer observed a strain of corn which bears in six weeks. Throughout the Colony corn is planted too far apart, and too many seeds are placed in the hole.... The time for planting is the month of May, or early in June, according to when the rains arrive. The Indians have a traditional "last date for planting," which frequently leads them into trouble if the May showers are late in coming.

Aspinall's text, with its alternating description and criticism of this system, begs many questions. Given the limits imposed by the text – a small pamphlet intended for wide distribution – why did Aspinall provide such a detailed description about maize farming? What makes recording these details important for a text that represents "British Honduras"? Why does he combine such description with criticism? What explains the noticeable shift in tone from that of Morris's resource assessment of the colony only forty years earlier?

Although both Morris and Aspinall were interested in defining the colony's resources and promoting capitalist investment, they write on different sides of a discursive shift. Both would agree that the methods of agriculture production are "primitive," but what this means has changed: for Aspinall, the "Indians" who are "the principal corn growers" produce "very satisfactory" yields, "considering the primitive methods of production." But these methods of production now constitute a *system*. Aspinall's text stresses this. There is only one footnote in the three pamphlets on British Honduras – it concerns, of all things, the Maya farm system (*"the 'milpa'* system"). The footnote reads: "The milpa system consists in the felling and burning of forest land and the crude planting of the corn amongst the logs and stumps with a pointed stick. After the crop is reaped the land is abandoned."[32]

At this point we can begin to delineate the rules of formation of the discourse about Maya farming. By reading the texts that have produced the Maya farm system, we can delineate six fundamental elements of this discourse.

First: Maya farming constitutes a *system*. The Maya farm and farmer are organically connected within a system that includes human and non-human actors. Three examples illustrate this element: Cook's influential article from 1919, entitled "Milpa agriculture: a primitive tropical system," explicitly defines the Maya *milpa* system as a "system."[33] Lundell frequently uses the concept of "system" throughout his important 1954 essay on "The agriculture of the Maya."[34] And as have we have seen from the appendix to the 1999 IFAD-GOB loan agreement, this continues to be of great importance.

But what exactly constitutes a system? From the OED we find that a system is "an organized or connected group of objects"; "a set or assemblage of things connected . . . so as to form a complex unity"; in scientific uses, a system is "a group, set, or aggregate of things . . . forming a connected or complex whole."[35] What is emphasized in speaking of the Maya farm system as a system is precisely this notion that its elements (nature, culture, labor, food) are articulated in a way that figures the entire complex as a unity, a totality. In figuring the system as a whole, Aspinall's language anticipates the desire to treat Maya farming "holistically."

To be defined as a system, an ensemble of things – objects, relations, processes, and connections – must be framed as a unity, a totality. At the level of representation, systematizing such processes and relations has three effects: first, establishing *boundaries* around the phenomena described (by marking off a set of processes or practices as "within" the system and differentiating these from things that are "outside" the system); second, *defining* the elements of the system and showing how they are related or connected (through concepts such as "level," "feedback," "loop," "flows," etc.); and third, *naming* the system (e.g., the "economic system"; the "political system"). These processes allow one to see the world in a new way: by systematizing an ensemble of disparate things, the "systems analyst" can see how its elements and processes function and connect. For instance, by defining "the farm" as a system, agroecosystem analysis provides the agronomist with a framework and a language to study how the interactions between crop and insect, soil and weed, and farm manager and state, are interlinked at different levels.

But once it is defined as a system, who can speak about farming? Can the farmer speak and be *heard* – except as an object, part of the system? These questions point towards one effect of systematizing discourse. Systems discourses shift our focus away from the particular, banal practices of everyday life and the existential forms in which we see things; they offer instead frameworks that fix meaning in useful ways. The systems approach thereby brings order to the scattered, fluid, and precisely unsystematic ways of speaking about and representing the world. The trick is that, in so doing, it presents the world as though that form of representation is only natural. But seeing and speaking of systems always involves positioning: systems discourse can only take place from a certain epistemic position – that presupposes a world of systems. Discourses on systems therefore stage the world in such a way as to make visible the unifying connections of social and natural life: but only to the technician, the systems analyst.[36]

Second: The Maya farm system is a *cultural* system. What defines the Maya farm system as such is not only its systematicity, but also its Mayaness, the relation between the system and the Maya qua ethnos. Whether one is examining *milpa* discourse concerning the "ancient" or "modern" Maya, the system is spoken of as constituted through cultural practices, defined in terms of *Maya* culture. Notwithstanding the facts that "milpa" is not a word from a Maya language (it is Nahuatl), and that the steel machete which is said to be instrumental was introduced in the 1500s, Maya culture is contained within, and produced through, the *milpa* system.[37] This reciprocal, essentialist reading of Maya agriculture has been refined, and reified, by anthropologists. A statement by Lundell captures this tone: "It may be safely stated that in general all the modern Maya employ the same system of agriculture as is found in Chan Laguna [where he studied], of which the essential features are the *milpas*, the semi-permanent plots, and the gardens behind the huts."[38]

Maya farming was one of the great topics of interest for postwar American anthropology, with labor sharing practices receiving specific attention.[39] Cook suggested that the Maya practice of collective planting is "a regular part of the system."[40] But, one might ask, what about those "Maya farmers" who go to their *milpa* alone? Or what happens in a Maya community where people decide to pay one another for their farm labor at the *milpa*?

Third: the Maya farm system is *primitive* and *inefficient*. The system is a low-technology, wasteful, unproductive way of producing a livelihood, as explained in a 1934 report for the Imperial Institute on the "Resources of British Honduras":

> The largest body of cultivators [in British Honduras] – the Indians – use no implements except the machete, or cutlass, and a pointed stick, while under their wasteful and destructive "milpa" system of cultivation they plant a new area every year, or at most every second year, by burning a new section of forest in the areas reserved for their use.[41]

Defined in these terms, the Maya farm system is treated as some kind of evolutionary mistake that defies a proper place in the order of things. It is so anachronistic, bereft of tools – Wright says that a Maya farmer's only tools are "a machete, a match, and a stick" – and utterly primitive that the Western mind is almost at a loss to understand it. Occasionally Western writers suggest that they are so appalled by the Maya farm system that they cannot write about it. When, in 1927, the Governor of British Honduras set out to describe the state of agriculture in the colony, he lamented: "The rural population has practically no conception of agriculture, other than shifting cultivation, mostly to supply the needs of the cultivator, of the crudest description." Shifting cultivation is so crude that he does not describe it. He adds: "This applies particularly to the Mayas and Spanish Indians."[42] Similarly, in his 1954 essay on the history of agriculture in British Honduras, Darcel writes:

> The Indians in the south are particularly primitive and have remained almost apart from outside influences. The system of "shifting cultivation" or the "milpa" system is practiced. The "bush" or secondary growth is cut and burnt and the seeds planted in holes made with a pointed stick between the charred roots. . . . Methods of cultivation are very primitive and have probably changed very little during the past one hundred years.[43]

What makes the Maya system so primitive, so crude? The lack of tools is frequently mentioned. Fire is troped as an elemental, as opposed to a refined, form of power. The absence of tillage, wages, and capital is treated as evidence of primitiveness. In short, the Maya farm system is not like the increasingly mechanized, and capitalized, forms of agriculture that were steadily displacing the peasantries of Europe and the United States in the early twentieth century.

Cook not only defines the MFS as a simple and low-technology system; he suggests that this may have been what prevented Western scientists before him from seeing it. "Doubtless the utter simplicity of the system has tended to keep it from being recognized or studied as a factor of tropical life."[44] Cook is correct: the Maya farm system could not have been "recognized or studied" by the West before this time because the "utter simplicity" of the landscape-and-"Indians" – that quasi-object we found described in the earlier texts – prevented its specification as a system, as an object at hand for rational description and willful development. It was *not seen*. By 1919 the conditions existed to make possible the visibility, and therefore the study, of the system.[45]

With the sighting, definition, and study of the Maya farm system, one had the tools needed to critique and improve the system. These are articulated in Cook's text. He argues that the primitiveness of the system makes it a good match for primitive people:

> Milpa agriculture appears well adapted to the needs of very primitive peoples, since only a minimum of labor and equipment is required. The ax or cutlass is the only tool that is necessary.... In typical milpa agriculture no labor is given to the working of the soil, either before or after planting. The crop simply is planted and allowed to grow.[46]

Agronomists would not write in such plainly racist terms today. Cook's argument, though challenged by a later generation of Mayanists – who would argue that Maya labor systematically and cooperatively produces their landscape – is deeply ingrained in the discourse of the Maya farm system.[47] In contemporary anthropology, one could not write "the crop simply is planted." But this does not imply that the Eurocentrism of this discourse has been eliminated.

Fourth: the Maya farm system is by its nature *destructive*. From the moment that Maya farming is spoken of *as* a system, writers define it as a process of destroying the forest or the soil. While the specific criticism shifts from text to text, the general pattern is to lament the fact that Maya farmers cut down the forest year after year and destroy the productive capacity of the thin tropical soils. Lundell says about this process: "The primitive *milpa* system of agriculture, as found in Chan Laguna, is effective where there is a limited population. Everything is taken from the soil, and nothing returned, so that *the system is by its nature destructive.*"[48] The same words appear throughout the discourse, even where the particular mechanism of destruction changes. Cook writes:

> Milpa agriculture as a system stands in contrast with tillage agriculture, in which plows or other implements are used to break the land before planting and crops are cultivated during the period of growth. From our standpoint of familiarity with tillage methods the milpa system appears not only temporary but highly destructive and self-limiting.[49]

This marks a clear shift from the earlier discourse about Maya agriculture, where the "Indians" were said to be living somewhere in the forest, enjoying the bounty of nature. From a discourse of haphazard and providential agriculture, we now find this concept of *milpa* farming as an essentially Maya system that, in its very work, destroys nature. This language used to describe this process often also sexualizes the landscape, for instance through expressions of concern for the "virgin rain forest." Cook's text says that Maya farmers "denude" the forest no less than eleven times.[50] (If the Maya farm system strips the land, who will stand up for the forest – to keep it properly clothed?)

The question of population is closely connected with this shift. At the very moment that this discourse defines the Maya farm system, it takes on a strong Malthusian tone: "the larger the population the more complete and extensive is the agricultural catastrophe which must ensue when a people who depend entirely upon the milpa system have exhausted their resources of production."[51] This conclusion is unqualified; the relationship is direct. Cook does not cite any data on population densities or land degradation to support this argument. Nor does this argument allow for strategies to mitigate against destruction, as Cook notes that Maya farmers employ practices to prevent overburning, but judges them irrelevant: "the natural limitations of the milpa system would be reached eventually." Given this inevitable conclusion, what is to be done? "Population would need to be restricted as well as fires if a permanent balance were to be maintained."[52] The MFS discourse thereby articulates the question of technology ("primitivity") with productivity, and grounds these together with population. The conceptual hinge for these points is *carrying capacity*, a concept that describes the number of people who could possibly be sustained with a given land area, form of technology, and level of productivity. Carrying capacity partly defines the MFS discourse as systematic, since carrying capacity is always measured in terms of a given system; one cannot speak of improving the efficiency of a system until it is established as such. The same is true of a system's relative sustainability or "permanence":

> With milpa agriculture the question of permanence hinges entirely on whether there are many people or only a few.... The essential inferiority of the milpa system lies not so much in its lack of permanence, since this would be secured if a proper balance of the population were maintained, but in the fact that the carrying capacity of any region must remain very small.[53]

What is written and rewritten, recursively, in this discourse – whether the element before us is the question of destroying forest, "denuding" the land, or the desire for securing a proper balance – is the solicitation of trusteeship. A system out of balance cries out for management. How this management is to occur, or who could be the trustee of the system: these questions are not yet posed explicitly in the 1920s.

Fifth: the Maya system leads to the *abandonment* of the land. The discourse on the MFS is replete with such references to abandonment. In his aforementioned essay, Darcel writes that "the 'milpa' is abandoned after its second year." What is curious is that the object that is abandoned by the essential "Maya" subject shifts in the discourse. For Darcel, it is "the 'milpa'" itself. In other texts, the land is abandoned; elsewhere it is a question of abandoning the great civilization of the ancient Maya (for instance, Cooke's 1931 article on the decline of the Maya civilization is entitled "Why the Mayan cities of the Petén District, Guatemala, were *abandoned*"). Other texts write that present-day, "living" Maya routinely "abandon" the land. Remember Aspinall's definition of the *milpa* system: "The *milpa* system consists in the felling and burning of forest land and the crude planting of the corn.... After the crop is reaped the land is *abandoned*."[54] This trope is found in many crucial texts on the MFS, including those by Cook, Lundell, Darcel, Cowgill, and others.[55] It leads to the abandonment of the land, of settled living, and of civilization.

"Abandonment" has several meanings which are linked in the discourse about the MFS. According to the OED, to abandon is "to forsake, to leave, to desert (a place, person, habit, practice, pursuit); to cease to hold, use or practice." Clearly these meanings are at work in the statements that the Maya are abandoning their farms. The Maya farmers leave their *milpas*; they cease to practice farming there. But "to abandon" carries a stronger connotation than "to leave," or "to move on," and it is this additional meaning, I argue, that explains its frequent appearance in these texts. Partly this meaning comes from the moral weight associated with "abandoning": one says that a parent abandons their child, and that a soldier abandons an army. One may leave a place, but one *abandons* something essential, something tied to one's being. We have already seen that the Mayanist discourse presumes the Maya to be essentially tied to the *milpa*. The moral weight of this discourse lies here: in the smooth, empirical description of the farm system, we find the Maya farmer abandoning that which is most essential to him. The MFS discourse thus produces a subject, the "Maya farmer," that is contradictory: he abandons his essence, that which constitutes him.

The OED gives as a second distinct definition of "abandon" one that is closely bound up with the first, and that applies particularly here, since it applies specifically to *land*. To abandon also means "to give up absolutely"; "to give up to the control or discretion of another"; "to let go"; and in the case of insured property, "to relinquish to underwriters."[56] In other words, when it is said that the Maya abandon their land, we should read this as a judgment: not simply that the Maya are *leaving* the farm, but that in so doing they are implicitly inviting another to take discretion for what they abandon. They walk away from the farm, as it were, with reckless abandon: they do so and relinquish claims to control the land. Remember the last words of Aspinall's footnote from the pamphlet promoting the Empire: "After the crop is reaped the land is abandoned." We cannot know whether this is written to solicit particular concerns for the land, or to facilitate scientific knowledge of the system, or to invite colonial settlement. These meanings coinhabit the sign "abandon," and the effects of this discourse articulate the meanings whether intended by the speakers or not.[57]

The sixth element of this discourse is that the Maya farm system produces a particular *spatiality*, namely, that of the scattered, impermanent, uncivil *alkilos*. We have seen in the texts by Miranda, Cockburn, and Crowe a pattern of describing "Indian townships" as "scattered and irregular."[58] With the shift to the discourse of the Maya farm system, this element becomes explained in a more rigorous way that links this to the other elements: because the Maya habitually destroy and abandon the land, they are forced to continually migrate:

> Although agriculture is usually considered a settled existence, . . . milpa agriculture is in a sense nomadic, from the need of moving about in order to find lands suitable for planting. Like wandering shepherds, the same

tribe might come back after decades or centuries to reoccupy a region that their forefathers had deforested and abandoned.[59]

The Maya farm system is therefore an exception to the rule: unlike the solid, stable, rural communities imagined to underpin European communities, this system is nomadic and scattered. It leads to wandering in the wilderness, an ungoverned and abandoned existence.

We should pause to reflect upon the shift that has occurred from the earlier Spanish discourse. The earlier ambiguous relationship between "Indians" and their "nature" has clearly given way to a position where the humanity of the Maya has been established. The Maya exist now, as Maya, as human beings named by their relations with a cultural group; and they are now a "population" that stands in apart from and in a separate relation to nature. This separation, this framing of the Maya as a culture group-population, occurs at the same time that the forest is seen anew – as a resource to be managed. Having once been scattered *in* the forest, practically *of* the forest, the Maya are no longer mere "forest Indians"; they are people represented as destroying the forest and abandoning the land. The Maya and the forest have been produced as separate things – separated in such a way that the relationship calls for trusteeship.

We should not be surprised to find, from Cook's text to the present, frequent calls to settle the Maya. The 1936 Annual Report of the Forest Department expresses the hope that, with the growing capacity of the Forest and Agriculture departments, "progress can now be made towards the betterment and *settlement* of the Maya Indian agriculture."[60] Two decades later, a soil scientist charged with drawing up the colonial government's land use plans will sum up the state's strategies for the Maya farm system with the words: "pin down the wandering *milpa*"[61] (more on this in chapters 4 and 5). Another two decades later, a study of the farmers of southern Belize concludes:

> Not only do farmers move from plot to plot, but they also abandon villages, and move to new areas from time to time.... A fundamental aspect in improving food production among many Milpa farmers, is the need to change them not only from shifting cultivation, but also from shifting their villages.[62]

Why are the Maya so shifty? How are we to stop the Maya from abandoning their villages? Permanent agriculture is proposed as an alternative to predatory nomadism. What is at stake is the proper way

to speak of the relationship between an agricultural system and movement, permanence, population, and civilization. The discourse on the
Maya farm system contains within it the instruments for reforming the
system, but they are within it, not represented as tools, logics, or concepts. They work through distinctions that stabilize and settle meanings.

Conclusion

> Prefer what is positive and multiple: difference over uniformity, flows over
> unities, mobile arrangements over systems. Believe that what is productive
> is not sedentary but nomadic.
>
> Michel Foucault (1983: xiii)

In the nineteenth century, people did not speak of the Maya farm system.
By the middle of the twentieth century it had acquired an ontological
fixity that is still strong. The discourse did not arise first in a Maya
language as an expression of an essentially "Maya" way of being; nor
did it emerge through the general application of the "farm system" or
"agroecosystem" approach (as we know it today) to the "Maya
world."[63] Nor did a set of statements that were dispersed suddenly
became systematic because of the agency of the "Maya." Rather, we
must conclude that this discourse came, epistemologically if not geographically, from the West. The Maya system was *made*. It is a work of
fiction – not untrue, but manufactured. Ontologically, the Maya farm
system did not precede its discursive production. How are we to explain
its existence? A change in the discursive formation created a new object –
one which has become fundamental to the way Mayanists speak of
what it means to be "Maya." Therefore, we cannot say that the key
texts merely *describe* the Maya farm system; they are what produce this
system. And they do so in a way that constitutes it as a system that
requires improvement and trusteeship: in a word, the system calls for
development. Here is the central finding of our archaeology: the Maya
farm system was made to be developed.

But this conclusion only begs the question: what *form*, what *kind* of
development is called for? Indeed, is this determined in advance? Are
there positive ways we could imagine this development unfolding – ways
that empower the Maya, that embrace their indigenous knowledge, their
market opportunities, and so forth? Is there no practical benefit to
describing, in careful ethnographic and agronomic detail, the work of

the Maya farm system – so that it can be integrated with capitalism in a relatively productive and harmless way? These questions frame Part II, where we will dwell upon the aporias of the Maya farm system's relation to capitalism qua development. But by way of underlining the pertinence of these questions within the discourse, let us return to the conclusion of Cook's essay. Asserting that "to recognize the limitations of this primitive system is not without practical bearing," Cook explains:

> Systems of agriculture that do not maintain the fertility of the soil are essentially nomadic and predatory. More permanent agriculture and more rational distributions of population are problems to be faced. Agriculture is the root of civilization, and the plant withers if the root decays.[64]

To end the predation; to end the nomadism; ultimately, to save civilization: that is the promise of the discourse on the Maya farm system.

The *milpa* and the Menchú Controversy

We should not take comfort in the thought that essentialist, Mayanist ideas about the Maya farm system are consigned to history. Insofar as there is a past, one does not need to look far to find it in contemporary Maya anthropology, where colonial discourses propagate with a vengeance. It would be irresponsible not to consider the best-known example: David Stoll's *Rigoberta Menchú and the story of all poor Guatemalans*. In this book-length critique of the veracity of certain statements by Rigoberta Menchú, Stoll offers a theory about the causes of the problems faced by the Maya of Guatemala. As opposed to most scholars who have studied the history of this war-torn and divided land, Stoll argues that the Mayas' problems do not stem from conquest, state violence, the expropriation of land, forced migration, and racism.[65] Rather, Stoll's thesis is that the problems of the Mayas in Guatemala are grounded in

> a degenerative process of population growth, slash-and-burn agriculture, and migration that is complicated, *but not altered in any fundamental sense*, by the Latino-indigena conflict and inequitable land tenure to which Rigoberta gives so much attention.[66]

Stoll is too clever to deny that the legacies of colonialism are involved with contemporary suffering. But he finesses these issues with the phrase "but not altered in any fundamental sense," which reduces the question of politics to two issues: "the Latino-indigena conflict" (as if this could be singular) and land tenure. He can then cleanly separate these two issues from population, agriculture, and migration, which are treated as apparently unmoving historical dynamics. Thus, Stoll circumscribes the debate to one about primary and secondary causes (which comes first, land conflicts or population growth?), with Stoll arguing in a neo-Malthusian vein that the Mayas are essentially to blame for their problems. They have too many children, cut down all the forests, and move itinerantly. They embody a degenerative process.[67] Anticipating the argument to come, we can see here why the Maya farm system is the hinge that connects colonial and developmental hegemonies: it articulates the question of civilization, and invites trusteeship, in a discourse with apparently objective, even scientific, qualities.

Since his thesis about the problems of the Maya of Guatemala is offered without any substantive evidence, Stoll is anxious to shift attention away from the question of how he could support such a claim (here I should remind the reader that Stoll claims he wrote this book to stand up for empirical truth). The sentence that follows the one I have just cited reads: "Romanticizing peasants is a hoary tradition that has the virtue of dramatizing their right to the land." Stoll's accusation here is that those of us who call into question the thesis that Maya reproductive, agricultural, and migratory practices are essentially "degenerative" are merely romanticizing the ugly facts of indigenous peasant livelihoods. That romantic views of peasant life have an ambivalent tradition is not a new claim.[68] The twist here – the argument Stoll proposes – is that Rigoberta Menchú's statements about her life merely romanticize indigenous struggles in order to win land claims, that is, to interrupt the colonial territorialization of Guatemala. In other words, he suggests that romanticism and indigenous movements are partners in crime. This should draw our attention to what is at stake in the discussion about the Maya farm system: that there are multiple approaches to Mayanism, and each is *in some sense* always already implicated with colonial tropes, including romantic portrayals of the Maya. The question, then, is which strategies of anti-racism and anti-essentialism offer the greatest room for maneuver. Stoll's angle is that of the colonial journalist-ethnographer who discovers the truth about the Maya through a positivist, empirical approach. With his theory about the roots of conflict in Guatemala, Stoll proves that these discourses still hinge on representations of Maya agricultural practices. His insinuation that "slash-and-burn" underlies

the Mayas' problems only underscores our need to reread the archaeology of the Maya farm system.

Some would challenge his approach on the grounds that it fails to
consider the ways in which the Maya are stewards of a culture that is
essentially sustainable and just. Fortunately these are not our only
alternatives, and we should reject both of the dominant representations
of Maya livelihoods, the destructive and the romantic. As we will see,
the latter too rests upon a colonial epistemology that treats "the Maya"
as an ethnos with essentialist (if superficially "positive") qualities. This
chapter has opened a postcolonial reading, one that begins to space
Mayanist discourses and scatter the systems approach that presently
settles them. This requires rejecting the systems approach – a settler's
discourse, one that resolves too much. This conclusion may seem to
contradict the spirit of Foucault's archaeological approach, which is
nothing but a systematizing discourse about discursive systems. But to
work in Foucault's wake, one must resist his often overly systematizing
tone. In reading the Maya farm systems discourse and the way it has been
mobilized for capitalism qua development (as I aim to do in Part II), we
should heed Foucault's remark that "what is productive is not sedentary
but nomadic."

It is Stoll's imbalance in his treatment of the standards of evidence
(high for Menchú, low for himself), and absence of critical reflection
upon his theoretical premises, that make his book about – rather, his
book *at* – Rigoberta Menchú such an "act of great and unjustified
aggression."[69] By combining his study of Menchú family history with
extremely broad claims about the Maya, he jumps scales of interpretation in great leaps, always landing with the confidence to state the facts.
Eduardo Galeano's criticism of Stoll captures the violence of such
anthropology:

> All of a sudden, the voices that speak of a scandal have multiplied, the
> voices that call Rigoberta a liar and that, in passing, repudiate the indi-
> genous resistance movement that she represents and symbolizes. With
> suspicious celerity, a smoke screen is rising to hide forty years of tragedy
> in Guatemala, magically reduced to a guerrilla provocation and to family
> quarrels, those typical "Indian things."[70]

This chapter has aimed at arguing that there are no such typical "Indian
things." Not even in the *milpa*.

Notes

1 Foucault, "The birth of a world." This quotation is from an interview conducted May 3, 1969.

2 IFAD and GOB, 1999. The remainder of Appendix V describes, in an explicitly anthropological tone, the "ethnic" composition of the region and the ways Maya farmers decide where to plant.

3 Note that the system here is referred to not as *milpa*, but as "shifting slash-and-burn." CARD is targeting the rural Maya not only because they are, on the average, the poorest people in southern Belize, but also because of this effect of "their system."

4 Foucault (1972: 131).

5 Ibid., p. 3, italics mine.

6 Few pre-twentieth century sources about the MFS are available in English because (1) the British did not found a formal colonial state in British Honduras until 1862, and (2) frequent fires and hurricanes have destroyed many records. Much of what is available comes from Spanish sources. A fuller study of Spanish colonial documents will be necessary to elaborate on my description of this archive. However, an exhaustive account of this archive is not necessary to examine the discursive shift entailed with the making of the MFS.

7 That is, the statements are structured in a way that provides them with a coherency that links statements, including the objects that are spoken of, with a more general system of possible statements.

8 Gallego y Cadena (1962: 142–3). My translation.

9 On the discourse of the abundance of nature in the Americas, see Lawson (1992).

10 One passage describes his time in Cahabon where the "Indians" were ill; therefore they could not "clean" or "watch over" their *milpas*. Describing one person in the community, a Doctor Sedoño, Gallego writes, "he could not clean or maintain his milpas, and therefore did not harvest this year; his soil is barren of corn and for this reason and because he does not have a region from which to continuously collect corn he suffers many hardships; Doctor Sedoño, who was counted upon by the village, finds himself in this pestilence" (my translation). Many Q'eqchi'-speaking Mayas of southern Belize trace ancestors to Cahabon, a mainly Q'eqchi' town in Guatemala.

11 Escobar (1841: 85–7). A parallel sentiment can be found in Lieutenant Cook's statement from 1769 which says nothing about Maya agriculture *per se*, but only that "their principle employment is the cultivation of the plantations" (1935: 34).

12 Here we can discern a new element of this discourse. Escobar notes that the space in which this maize is grown in a natural, spontaneous way, has been "*anciently inhabited* by the Indians." This marks a discursive encounter

between "Indian" agriculture and the "discovery" that the farmers are descendents of "the ancient Maya" – a relation of great importance to Mayanism (see chapter 3).

13 Similarly, Morley cites Bishop Landa to the effect that: "In cultivating the land they do nothing more than clear the brush, and burn it in order to sow it afterward, and from the middle of January to April they work it and then when the rains come they plant it, which they do by carrying a small sack on the shoulders, and with a pointed stick, they make a hole in the ground, dropping in it five or six grains, covering them with the same stick. And when it rains, it is marvelous how it grows" (de Landa, *Relacion*, cited in Morley 1946: 147).

14 Miranda (1955), cited in Reina and Hill (1980: 77).

15 Miranda (1955), cited in Reina and Hill (1980), my translation. A demonstration of the effect of the shift in discourse comes from the interpretation of this statement by Reina and Hill: Miranda, for them, describes "a delicate and complex system for the provisioning of Maya people in the transitional tropical highland-lowland zone of the Alta Verapaz" (78). Miranda, in fact, does not describe a complex system. The concept and discourse for such a description does not exist; for Reina and Hill's reading, it is fundamental.

16 Artiga, 1699, cited in Jones (1982).

17 This also explains why, though I rely here on Spanish sources, I am not simply describing the Spanish "worldview." One could ask: why attribute this "rule" to discourse and not to an essentially Spanish way of seeing "Indians" as part of nature? The answer is that we are trying to define the rules of this discourse without appealing to a broader transhistorical structure – especially of an anthropological type (e.g., "the Spanish worldview"). Such a structure would not only prove unwieldy but would only beg the question of its own archaeology.

18 Miranda 1953, p. 355, cited in Reina and Hill (1980).

19 I do not equate sixteenth-century Spanish writing on the Maya with nineteenth-century British texts. There are important differences: the meaning and purpose of empire in Central America; the nature of a Maya person: all this fundamentally and decisively changed in the intervening three hundred years. Nonetheless, the parallels on the question of the spatiality of houses and agriculture are striking. And it is important to note that neither Miranda nor Crowe treat Maya farming as something that is systematic or a thing that could be abstracted from the very existence of the people (or souls) that are encountered.

20 Crowe (1850: 43–4).

21 Now known as Corozal, in 1875 this was the major town of northern British Honduras, inhabited mainly by Yucatec Maya-speaking farmers.

22 Cockburn (1875: 26).

23 Ibid., p. 28. Note the double meaning of "want" here: that which is absent, and that which is desired. Similar statements can be found in the writings of

Tozzer, the first widely recognized Maya ethnographer. In his study of the Mayas and the Lacandones (1907), he says that "agriculture is necessarily very crude among both the Mayas and the Lacandones, owing to the nature of the ground and the lack of modern tools and methods" (51). Although Tozzer's ethnographic tone clearly differentiates him from earlier writers, he does not describe Maya agriculture as a system; the other elements of the ethnographic MFS discourse (discussed below) are not present, either.

24   Seymour writes: "The population of British Honduras may be divided into four classes[:] 1st the Maya and Chichenga Indians (Aztecs). 2nd Spanish refugees from the neighboring republics who already form the numerical majority. 3rd White and black creoles, Africans and all who speak the English language. 4th Caribs" (Seymour 1859, cited in Burdon 1935: 221). This particular set of racial distinctions (called "classes") is interesting for several reasons. First, it groups "the Maya" as a unit (despite the fact that there is no singular "Maya group") with the "Chichenga Indians," a group that falls out of racial discourse in British Honduras by the early twentieth century. Second, these categories exclude Europeans, including the British (who either do not have a race, or did not see themselves as residents of British Honduras). Third, although the British are not mentioned here, all those who speak English are combined into one class ("all who speak the English language"). This schema is loosely based on language families. On this basis, "Caribs" (the Garifuna) are separated from "Africans." The distinction between the Garifuna and the Creoles was fundamental to colonial hegemony in nineteenth-century Belize; consent of the Creole working class was won partly through practices that defined them as distinct from and superior to the Garifuna and Maya. The discursive and social separation was coordinated with a strict spatial separation that prevented Garifuna people from living north of Dangriga (or Stann Creek Town).

25   Ibid. Note the two references to "civilization" in the passage, which express an anxiety that the Maya are missing out on the effects of civilization because they are shuttling between the forest and the town. I believe the last two sentences of this passage refer to the Mayas on the southern periphery of the colony.

26   Ibid.

27   Gibbs (1883: 162).

28   Morris (1883).

29   To return briefly to our discussion about territorialization from the Introduction and chapter 1: in the section on the geography of the colony, Aspinall writes that "The greater part of the interior of British Honduras is still unexplored, and the western boundary is an artificial one, of which only part has been surveyed" (1924: 9). We should note here the artificiality and lateness of boundaries, and also the anxiety about consolidating the colonial geography of British Honduras.

30 "Corn" is an old English word that refers to the locally dominant cereal grain of a region; thus in different parts of England and Scotland, "corn" signified wheat, oats, or barley. Its use as a term for *Zea maize* started in the United States, where the word "Indian" was eventually dropped from the term used to describe *Z. maize*, the dominant cereal: "Indian corn." "Maize" comes from the Spanish *maíz*, an adaptation of an Aztec word. None of these words, in short, are rooted in Maya languages. The Q'eqchi' term is *ixim*.

31 Ibid., pp. 9–10. The "*" is a footnote marker in the original text; I discuss the footnote below.

32 Ibid., p. 9.

33 Cook (1919).

34 Lundell (1933: 67, 69, 70, 72, 73, 74, and 77).

35 OED, second edition, Vol. 17, p. 496.

36 Timothy Mitchell makes a parallel argument in *Colonising Egypt* (1988) where he explains how it is possible that Egyptologists can *see* and analyze the effects of Egyptian culture qua system – while the Egyptian other is simply a "part" of the cultural system.

37 Like "Maya", "milpa" is a word that was taken up and distributed by the Spanish. A Nahuatl word used to mean "maize fields," by the 1920s it was taken to refer to the Maya agricultural system. Cook writes:

> English and other European languages have had no recognized names for this primitive system of crop production which is general in hot countries.... Milpa agriculture would be a convenient designation, the native work "milpa" having been adopted by the Spanish-speaking people of Central America in the sense of a maize field, or a clearing in the forest, cut and burned for planting maize. As an Aztec word, milpa is derived in Robelo's Diccionario de Aztequismos from "milli," a planting, and "pa," in, with the remark: "Now applied only to maize." The vocabulary of Brinton's Maya Chronicles includes a verb "mulba," "to congregate, to come together," the possible connection being that all the people of a community usually work together in cutting and especially in planting a milpa. (Cook 1919: 308)

Elsewhere in this text, Cook gestures to the fact that Maya speakers do not speak in Maya about farming in their languages as a system. He writes, "Some families might go farther out to 'make milpas' and carry in their corn..." (p. 315). The scare-quotes around the expression "make milpas" suggest that, in the speech of his native informants, cornfields are places that are produced. In conversations with Maya farmers in English or Spanish, Maya speakers tend to refer to cornfields as *milpas*; but this is not the case when speaking in Maya languages. Not only are other words used to speak of the place where corn is grown, but the expressions related to farming ("go to the farm," etc.) are not translatable in terms of a "farm system."

38   Lundell (1933: 69). I do not need to belabor my readers by citing all of the
     examples that disprove Lundell. The point is not that Lundell is wrong, but
     that such statements define this discourse. For studies on the diverse
     agricultural practices that have sustained Maya communities, see Harrison
     and Turner (1978); Wilk (1981, 1985a, 1997).

39   There are no less than five different ethnographies of the Q'eqchi' of
     southern Belize from this period that examine agricultural labor: Rambo
     (1962); McCaffrey (1967); Howard (1974); Wilk (1981); and Berte (1983).
     Wilk says somewhere that in the 1980s there was a Peace Corps volunteer
     for every 1,000 Belizeans. Over the past four decades, roughly the same
     ratio holds for anthropologists and the Maya in southern Belize.

40   Cook (1919: 308).

41   Anon., 1934. "The Resources of British Honduras", p. 383.

42   Burdon (1927: 35).

43   Darcel (1954: 3).

44   Cook (1919: 308).

45   This is not to say that the word "system" was never used before 1919 in
     relation to the Maya. The earliest reference to the Maya agricultural system
     that I have found – one that does not elaborate on the way the system
     works in the sense of Cook or later agronomists – comes from Stephens'
     1843 *Incidents of travel in Yucatan*, an immensely popular book and one of
     the founding texts of Mayanism. Stephens does not miss the opportunity to
     stress the simplicity of the system:

     > The system of agriculture in Yucatan is rather primitive. Besides hemp and
     > sugar, which the Indians seldom attempt to raise on their own account,
     > the principal products of the country are corn, beans, and calbazas,...
     > camotes,... and chili or pepper.... Indian corn, however, is the great staple,
     > and the cultivation of this probably differs little now from the system
     > followed by the Indians before the conquest.... The hoe, plough, and harrow
     > are entirely unknown;... the machete is the only instrument used. (Stephens
     > 1963: 136–7)

     Since the MFS comes into existence as a discursive object through the work
     of Mayanism as a form of disciplinary knowledge, it is appropriate that the
     term "system" is first used by Stephens. By suggesting that the system
     "probably differs little now from... before the conquest," Stephens poses
     a question that will structure the discourse on the Maya farm system
     throughout the twentieth century – as we will see in the following chapter.

46   Cook (1919: 308). This paper may well be the first to articulate the
     different elements of MFS discourse. Yet, whether or not this particular
     text is the first is not critical for my argument. Cook's article is significant
     because it defines and names the Maya "milpa system" (see note 37,
     above), and because it became an obligatory passage point for the field
     of "milpa studies" that has followed (on "obligatory passage points" of
     scientific discourses, see Latour 1987).

47 Agronomists and anthropologists have challenged the "primitivist" aspects of the MFS discourse in Belize (cf. Wilk 1981, 1997). But these texts still turn within an ethnographic mode of representing that produces the Maya as other. Although the "simplicity" of the Maya system is challenged, and although the system is seen as having positive values, the structure of the discourse remains.

48 Lundell (1933: 67).

49 Cook (1919: 323).

50 Ibid., pp. 310, 311 (2), 314, 315, 316, 319 (2), 322, 325, 326. For instance, the "repeated burning of the forest for agricultural purposes" produces a "state of complete denudation and exhaustion of the soil."

51 Ibid., p. 314.

52 Ibid., p. 318. Cook's text does not suggest the means to find, or maintain, this "permanent balance."

53 Ibid., p. 323.

54 Aspinall (1924: 9), italics mine.

55 See, for instance, Cook (1919: 308, 309, 318); Lundell (1933: 67, 69); Darcel (1954: 3); Cowgill (1962: 276).

56 OED Vol. 1, 2nd edition, pp. 9–10.

57 And not only these. The verb "to abandon" has two obsolete meanings that could also be said to be at work here, however buried they may be: "to subjugate absolutely" and "to banish." Both stem from the old French "à bandon," i.e., of the staff, of the command of the King. On abandonment, the sovereign ban, and the law, see Agamben (1998).

58 Crowe (1850).

59 Cook (1919: 312). We should note that the description of the Mayas as a "nomadic" people is usually qualified, as it is here ("*in a sense* nomadic"). Similarly, the Assistant Magistrate of the Toledo District in 1967 writes that "The Indians are semi-nomadic, and their whole social living is oriented around this type of life.... As a result of this semi-nomadic way of life, not much good use is made of the land" (Flowers, AB MC 5325, p. 5). He does not say what constitutes "semi-nomadism." Why is nomadism referenced so frequently within this discourse, but qualified? Second: consider the biblical trope of Cook's portrayal of Maya farms as "wandering shepherds," a lost "tribe that might come back." Are the Mayas the lost tribe of Israel after all? Who watches over the wandering Maya?

60 Anon. (1937).

61 Wright et al. (1959). See chapter 4.

62 Caribbean Food and Nutrition Institute (1979: 6). The CFNI report attributes the *milpa* system not only to "Maya Indians" but also to "East Indians... and Caribes [sic]." In their view, all the farming groups of southern Belize (except for the Mennonites) need integrated rural development on these grounds.

63 The discourse about the Maya farm system emerged several decades earlier than agroecosystem analysis. The contemporary concern for "holistic" agroecological analyses of the "farm system" emerged in the 1970s at the intersection of agronomy and systems ecology. The MFS is not simply a derivative of this agronomy-based discourse. We cannot explain the emergence of the MFS in terms of the general encroachment of the "farm system" concept from centers of learning in the United States.

64 Cook (1919: 326). Elsewhere Cook suggests that this lack of balance between population and food has destroyed ancient civilizations, including the Maya: "The general tendency of civilization is to develop large centers of population without corresponding improvement of food supplies, so that the practical limits are reached" (315).

65 On Guatemala's history as three iterations of conquest, see Lovell (1988). The third round of conquest continues, even after the 1996 Peace Accords. On the ongoing theft of land from Maya peasants in Guatemala, see Black (1998).

66 Stoll (1999: 19), my italics. For Stoll's argument on the political ecology of "what happens between Indians," see pp. 15–27.

67 Note that the "gen" in de*gen*erative and indi*gen*ous is the same: the politics of birth here connect the question of infantilization and the need for development. Note too that Stoll's "degenerative process" thesis recapitulates the archaeology of the Maya as *fallen* original Americans.

68 It was the question of the very possibility of representation of peasants that led Marx to his famous conclusion that "they cannot represent themselves; they must be represented."

69 Pratt (2001: 29).

70 Galeano (2001: 149). A quibble: as Galeano knows, Stoll's book obscures not 40 but 500 years of tragedy.

3

An Archaeology
of Mayanism

Colonialism and imperialism are accused of being sadistic forms of invasion and domination. But the most typical subversion of colonialism is its aestheticentrist way of appreciating and respecting the other.
Kojin Karatani (2000: 145)

In the annals of European history, most indigenous people are discovered once. The Maya were discovered twice. If we define these two moments in the arc of conventional European historiography, we could say: the Maya were discovered once at the beginning, and again near the end, of the age of empires. The Maya were first discovered by Spanish conquistadores and Catholic fathers, who met people and classed them "Maya" – and pronounced them potentially Christian even as they fought and killed them. This was the first of two Maya "discoveries," albeit one that carried on for centuries and at different rates across Mesoamerica (the last independent Maya kingdom, of the Itzaj', did not fall until 1700).[1] But in the mid-nineteenth century, toward the other end of the age of empire, the Maya were rediscovered as the sons and daughters of a great civilization. The second discovery of the Maya occurred under the banners of liberalism and science in the mid-nineteenth century, when the Maya became ancient in ways that other nations were not, and their importance for the human sciences became recognized. During the same period, their lands – those that had not already been expropriated or otherwise dispossessed – underwent a new round of conquest in Mexico, Belize, and especially Guatemala. The violence accompanying the second discovery was no less than the first. This new round of dispossession was accompanied by the emergence of a new way of knowing the Maya – who they were, what it was to be "Maya" – that is, in Karatani's terms, aestheticentrist. Following the patterns of use within the literature, I will refer to this way of knowing the Maya as "Mayanism." This way of knowing comprises my object of analysis in this chapter.

By Mayanism I refer to an institutionalized discourse concerned with discerning and explaining the nature of the Maya qua ethnos. In framing my argument in this way I am drawing deliberately from Edward Said's 1978 study, *Orientalism*, which examines the histories and effects of orientalist discourses and institutions. Said's critique of orientalism is arguably the first, and perhaps still the best, demonstration that our ways of understanding the world are produced by discourses that have histories bound up with the experience of European colonization. His study of orientalism examines how what he called "imaginative geographies" – effective representations of the world – are produced by colonial discourses and may have immense effects long after the end of formal colonialism. With this argument Said is not trying to turn the tables on the West (to suggest somehow that the Orient is truly better than the West), but rather to examine and explain the *production* of this very distinction in the imperial encounter – as well as political effects. As Said explains in the 1994 afterward to *Orientalism*, his aim is to extend "post-colonial concerns to the problems of geography... [and] to challenge the notion that difference implies hostility, a frozen reified set of opposed essences, and a whole adversarial knowledge built out of those things."[2] Again, the problem is not one of finding a balanced view of the "East" from the "West," but the very framing of the world in such terms.[3]

Although most readers have emphasized the importance of Foucault's approach for this project, a strong argument can be made that Said understated the influence of Gramsci on *Orientalism*. Reflecting on this period, Said noted in a 1993 interview that he drew from Gramsci a sensitivity for the ways that imperialism produces geographies that come to be taken as "natural":

> The single most important thing that I took from Gramsci... [is] the idea that everything, including civil society to begin with, but really the whole world, is organized according to geography. He thought in geographical terms, and the *Prison Notebooks* are a kind of map of modernity. They're not a history of modernity, but his notes really try to *place* everything... [In the notebooks] there was always some struggle going on over territory.[4]

Read in this light, Said's *Orientalism* can be read as an extended Gramscian geographical study of the practices through which large numbers of people become quite literally *oriented*. Orientalism, Said shows, produces "the Orient" as "natural geography," an object for description, while simultaneously disciplining knowledges about this object to conform to an imperial way of knowing it. Thus, "the Orient" is a settled place, a bounded region set within the real world, with a specific form of external and expert knowledge appropriate to it.

There are important limits and exclusions in Said's reading of Gramsci. To my knowledge, Said never dwells on Gramsci's communism, his novel approach to the state, or his strategies for achieving proletarian hegemony. Nor do Said's writings on Gramsci's "social groups" address the fact that, for Gramsci, understanding the existence and transformations of such groups requires a Marxist theory of class.[5] But the most substantive criticism of Said's application of Gramsci and Foucault, I would argue, is that his analysis of colonial discourses ironically reifies the very distinction that he tries to call into question, that Said reproduces the very difference between "the West and the rest" that he aims to critique. In his polemic against Said, Ahmad writes of *Orientalism*:

> It is rather remarkable how constantly and comfortably Said speaks ... of *a* Europe, or the West, as a self-identical, fixed being which has always had an essence and a project, an imagination and a will; and of the "Orient" as its object.... In other words, he seems to posit stable subject-object identities, as well as ontological and epistemological distinctions between the two.[6]

I think it is fair to say that Said invited this sort of critique when he summarized the basic lesson of *Orientalism* in these terms: "The general basis of Orientalist thought is an imaginative and yet drastically polarized geography dividing the world into two unequal parts, the larger, "different" one called the Orient, the other, also known as "our" world, called the Occident or the West."[7] The world is plainly divided, deeply and unequally divided, and orientalist discourses are only partly responsible for the spatial dispensations of power and violence.[8] And they are emphatically discourses in the plural, not one; the relations between "Europe" and its others have never been singular nor stable. From this we may conclude that Said's analysis of orientalism was simply wrong (this is the conservative and vulgar Marxist reading) or we may conclude that the modes of orientalist thought (as Said uses the term) must continue to be analyzed and criticized while attending to their heterogeneity and particularity. If we would like to retain the basic lesson of

Said's critique of orientalism – that Western colonialism was made possible by imperial ways of framing the world that live on, that contribute towards the reproduction of unequal geographies and the intense divisions between peoples – then we have to situate his critique, loosen it up, space it, and put it in play with the critique of capital.[9] I adopt this latter position because I feel that Said's critique of orientalism still has many lessons to teach.

We can see the relevance of this if we abide by the Maya. Reading through the colonial archive in Belize and London, it became clear to me that the way that the British colonial officers wrote of the Maya was at once "orientalist" in Said's sense, and yet also part of a distinctive regional discourse about the Maya – one that inherited key concepts, tropes, and words from the Spanish (who play little part in Said's study). Mayanism could be further construed as "orientalist" insofar as several of its key practitioners actually studied in Egypt, were inspired by orientalist texts, and applied orientalist practices in Mesoamerica. Yet we cannot get around the fact that Mayanism does not fit neatly under all the things Said has to say about orientalism. To begin with, the people that Mayanism knows do not live in "the Orient." Indeed, Mayanism emerges precisely around the recognition of the very unfamiliarity, indigeneity, and "Americanness" of the Maya – a triple connotation well captured in Mason's not atypical description of the "Maya world" as a "strange, purely American nation."[10] Mayanism emerged within nineteenth-century European thought after the discovery of Maya ruins, and the recognition that they were the work of "American" nations, brought explorers, collectors, and scientists from Europe and the USA to study the Maya.

It is the works of these experts – who often called themselves "Mayanists" – and not the Maya *per se* that are my concern in this chapter. Of course, my approach does not deny the existence of Maya people, nor the present-day "pan-Mayanist" movement.[11] But they are not my object: that again is solely Mayanism, the institutionalized Eurocentrist discourse that emerged in the nineteenth century to define, describe, and explain the Maya. My accomplices are texts written in this Mayanist tradition.[12] I emphasize some of the strongest works in this tradition because they have had the most power in defining European views of the Maya. And with a few notable exceptions, the most influential works have been written by European writers – often writing in Mesoamerica, geographically, though always, epistemologically, from Europe. I excavate Mayanism by investigating key sites: I begin with Bartolomé de Las Casas, the first important European advocate for indigenous rights – and the forerunner of Mayanism. I then turn to the writings of early Maya archaeologists – John Stephens, Fred Catherwood, and Thomas Gann – who traveled through the Maya region in the nineteenth century and wrote popular accounts of their explorations. Finally, to draw out the spatial adumbrations of Mayanism, I discuss two celebrated settings where the Maya have been territorialized: Miguel Ángel Asturias' monumental novel, *Hombres de Maíz*, and the British Museum. Given the massive volume of the Mayanist archive (the Maya may well have the dubious distinction of being the most-studied ethnos in the world), I cannot claim that these are "representative samples" of Mayanism. I focus on these texts and sites because they are among the richest, most influential works in the tradition; each marks a keystone in the archaeology of Mayanism.

Let me briefly preview the argument. The Mayanists were provoked by the mystery of the Maya: a great civilization grew in the New World tropics, then suddenly disappeared.[13] Mayanism emerged as a way to explain this mystery while also defining who the Maya were – to explain how to appreciate them aesthetically. Mayanism thereby made possible a way of speaking of and for the Maya without the Maya. In the resulting discourse, the Maya are framed as a fallen people: once great and worthy of admiration, they still receive appreciation as relics of an ancient and mysterious civilization. Through Mayanism, the Maya stand before the West as an allegory for the decline of our own civilization, and thus the living Maya are an especially pertinent site for our own moral investment. On this basis – in order to carry forward civilization in the face of its potential demise – the Mayanist discourse points the way forward for the Maya farm system: settling, territorialization, and capitalism qua development.

Mayanism, *de Vera Paz*

For all the peoples of the world are men, and the definition of all men, collectively and severally, is one: that *they are rational beings*. All possess understanding and volition, being formed in the image and likeness of God; all have the five exterior senses and the four interior senses, and are moved by the objects of these; all have natural capacity or faculties to understand and master the knowledge that they do not have; and this is true not only of those that are inclined toward good but those that by reason of their depraved customs are bad.... Thus all mankind is one, and *all men are alike in what concerns their creation and all natural things*, and no one is born enlightened.... And *the savage peoples of the earth may be compared to uncultivated soil* that readily brings forth weeds and useless thorns, but has within itself such *natural virtue* that by labor and cultivation it may be made to yield sound and healthful fruits.

Bartolomé de Las Casas[14]

The works of Bartolomé de Las Casas (1484–1566) have long inspired advocates for indigenous rights in the Americas. Las Casas is often said to have been the first to articulate the modern concepts of human rights, and credited with fundamentally changing the terms of debate in Europe on the humanity of the indigenous peoples of the Americas. His criticisms of Spanish colonial practices – described in all their brutality in his famous *Short Account of the Destruction of the Indies* (1542)[15] – were fundamental to the passing of the Spanish Crown's *Leyes nuevas* of 1542–4 and later attempts to mitigate the severity of imperial violence. Until relatively recently, the dominant interpretation of Las Casas in Spain was that his account of Spanish imperialism was unbalanced and that he did wrong to Spain by exaggerating the violence of conquest. This so-called "black legend" was exploited by the British and the US in their wars against Spain.[16] Contemporary evaluations of Las Casas advance more nuanced readings. I would endorse Hardt and Negri's assessment in *Empire* that Las Casas was one of a series of Western utopians who contested imperial power in ways that were at once ambitious and revolutionary, yet fundamentally limited. Comparing Las Casas to Toussaint L'Ouverture and Karl Marx, Hardt and Negri argue that the long "progression toward globalization" wrought by European colonialism has always been accompanied by "utopian tendencies... the love of differences[,] and the belief in the universal freedom and equality of humanity."[17] The pertinent question is, of course, *how* difference, freedom, and humanity are materialized in thought and practice. And here it is Las Casas' complexity – his greatness and his limits – that make him an important theorist of colonialism.

It is hard to mistake Las Casas' importance in Mesoamerica, as he appears so frequently in public art, paintings, statues, and murals in the churches and towns where he worked. Notably, most of these monuments date from the nineteenth century – well after the wars for independence from Spain – and Bolívar was not the only liberal to see in Las Casas an exemplar for the morality of American resistance to Spanish rule. Consider figure 3.1: Parra's 1875 realist portrait of Las Casas from Mexico's *Museo Nacional de Arte*. Las Casas is here depicted as the moral conscience of Europe, reflecting the light of reason while appealing to God from the scene of imperial atrocity. Note that Las Casas is visually framed by the hallmark of Mayanism, the Maya ruin. The painting allows us to appreciate at once the grandeur of the ancient Maya, the moral failure of the Spanish colonists, and the singularity of Las Casas. But note that the indigenous woman in the painting does not lament before the statue of the deity to her left, but towards Las Casas. I concur with Widdifield's reading that this is to imply the inevitability of conversion and assimilation: "The future is guaranteed not to the reproduction of the Indian" – since the female figure's male partner has been

Figure 3.1   Parra's portrait of Las Casas, 1875
Felix Parra, *Fray Bartolomé de Las Casas,* 1875. Oil on canvas, 141 x 105 inches. *Museo Nacional de Arte,* Mexico City

slain – "or to the continuity of native religion[,] but to the reproduction of Christianity."[18] Parra's painting therefore offers a means of aesthetically appreciating the greatness *as well as* the fall of the Maya – and all this, remarkably, while implicitly celebrating the redemption of the Maya through Christianity and European enlightenment (which will check colonial violence). In this it is a metaphor for Mayanism itself.

Although representations of Las Casas and his contributions to the history of American thought have undergone profound changes over the centuries, those who have emphasized his importance have stressed his status as a thinker of the Americas and the "moral conscience of the 'enterprise of the Indies.'"[19] When Yáñez concludes his study of Las Casas by claiming that he was "one of the great creators of the American *ethos*," he reiterates Simón Bolívar, who once called Las Casas "the Apostle of the Americas."[20] Although I depart from these characterizations, I underscore their troping of Las Casas as a contributor to something essentially *American*. Indeed, Las Casas' thought was intensely geographical. My aim is to draw this out, for it marks the ground of Mayanism.

To appreciate the geographical character of Las Casas' thought we must first grasp the essence of what I will call, following Alberto Moreiras, Lascasian hegemony.[21] Few thinkers can be said to have defined and also executed a particular form of hegemony; Las Casas is one. His contribution as a thinker lies not only in his defense of indigenous rights (which I take up below), but also his assertion that there is only one moral and effective way to attract people to the Christian faith. This argument, elaborated in a Latin manuscipt called *De unico vocationis modo* (hereafter, *The only way*) provides the fundamental theological basis for his argument for the rights of indigenous nations. In this text he argues that all people are potentially Christian, and that the church has a responsibility to preach to all peoples: "God's chosen should be called, should be culled from every tribe, every language, every corner of the world."[22] But, Las Casas avers, there is only one ethical, effective, and Christian way to execute this task: not through violence, as the early Spanish colonists claimed to do, enslaving indigenous people in order to save them, but through "the arts of persuasion [and] supplication." Las Casas argues that the ideal Christian missionary must bring the would-be Christian into the faith "by induction and smooth attraction, conduction, through the repetition of acts intended to engender in their heart certain customs and dispositions" which over time will lead the believer to feel "a natural inclination toward the truths."[23] Here Las Casas presents a theory of hegemony not unlike Gramsci's, but with an entirely different orientation. Like Gramsci, Las Casas focuses on the achievement of power through persuasion, built through iterated practices that transform customs and dispositions until the would-be Christian is naturally inclined to the truth. Whereas Gramsci's studies of hegemony examine the conditions needed for communist revolution, in Lascasian hegemony we see feudal Europe producing hegemonic imperial reason.[24] Thus, we cannot treat Las Casas as merely a liberator, a beneficent father of human rights. Fundamentally his aim was, in Moreiras' words, "to avoid the destruction of the imperial territories and their native inhabitants for the sake of *a more perfect territorialization.*"[25] Las Casas wished to make European colonialism hegemonic as a form of territorial power. Insofar as Lascasian hegemony always rests upon the possibility of a greater, more violent terror, one that advances alongside and sometimes before Las Casas' "arts of persuasion," it is always already an effect of terror. Thus Lascasian hegemony is more violent than it seems. This limit continues to work through the territorializing violence that places indigenous nations within a worldliness defined by and for the West. To see this we need only follow Las Casas' first steps in the Maya world.

The Tezulutlán Experiment

Las Casas' time among the Maya began with a capricious change of weather. En route from Peru to Mexico in 1531, Las Casas' ship was "becalmed" off the coast of Nicaragua and forced to take refuge. Las Casas took this opportunity to survey the state of colonization in the territory, and he wrote to the King on October 15, 1531:

> Beyond Lake Nicaragua, there live a number of tribes not yet subjected to Your Majesty. They are deadly enemies of the Christians because of the atrocities from which they, like the other Indians, have suffered. If Your Majesty sent us the appropriate *cédula*, we could pacify them and bring them to the service of Your Majesty, on condition that no Christian, great or small, may have anything to do with them, nor may they be subjected to any individual Christian under any kind of slavery.[26]

The requested *cédula* was sent in July 1537, but by the time it arrived in Nicaragua Las Casas had left for the center of Spanish power in Central America – Santiago, known today as Antigua, Guatemala – where he had been invited by the Bishop-elect Francisco Marroquín to put his ideas into practice. Upon arrival, Las Casas went to work "pacifying" the Mayas of Guatemala, but he encountered resistance from the Spanish colonists whose interests he threatened. These settlers proposed that he practice his ideas in the area called *Tezulutlán*, the "land of permanent war" – a region known today as the *Verapaz* of Guatemala, but also including much of the Peten and Belize. Tezulutlán gained its name by the inability of the Spanish to conquer the formidable Maya resistance. The challenge represented by the unconquered Mayas of Tezulutlán provided Las Casas with an opportunity to materialize, geographically, his argument that the only way to convert Christians was by persuasion. Las Casas negotiated an agreement with the Governor of Guatemala, who wrote to the Emperor of Spain on October 16, 1539:

> In [Tezulutlán] there is an extensive, warlike territory that has never been subdued.... Pedro Alvarado did not succeed in subjugating it during the time that he was here.... *Another method of obtaining this land* is being tried. Father Bartolomé de Las Casas...[is] succeeding in the *peaceful conquest* of this warlike territory. To this end [he has] been carrying on negotiations with the Indians, unknown to any Spaniard save themselves and me. The reason is that we are convinced that if the Spaniards knew about it one of them would do all in his power to upset these negotiations.... I hope with the aid of God to render thereby a great service to God and Your Majesty and put an end to the killing, robberies, and the like that usually accompany conquest.[27]

Las Casas and his assistants set to work to secure the peaceful conquest of Tezulutlán, which Hanke calls "the last great enterprise which engaged [Las Casas'] attention in America," one that "embodied his most important concept of the proper relationship between Spaniard and Indians," that is, that "Indians must be Christianized by peaceful means alone."[28] I will only briefly summarize this complex tale.[29] They began by writing poems and songs to depict the Christian story of creation and the fall from paradise and translating them into Q'eqchi'. (For this, Remesal suggests, "they were most likely the first men to compose poetry in the Indian language, the first Spaniards that is, and they should surely be remembered as such.")[30] They trained four Christian Mayas, traders who spoke the local Maya language, to perform the poems and songs, and sent them to Kob'an, the largest town in Tezulutlán. The traders carried local wares to sell, "but Las Casas additionally gave them Spanish scissors, knives, small mirrors and the bells which Indians liked too much."[31] The results were propitious: when the community heard the songs and stories, they invited the Spanish to send one of their padres to come and teach them more of

Christianity. In Santiago, Las Casas selected Father Luis Cáncer, who entered the territory of Kob'an, "received by triumphal arches." Cáncer preached for several days and won over the leaders of the Q'eqchi'. As Remesal dryly notes, he benefited in this work from a document he carried that was "signed by the king's representative in Guatemala which certified that no...native [would] be pressed into a Spaniard's service."[32] Alongside the teaching of Christ, then, Lascasian hegemony was won through a territorial agreement (complemented by "Spanish knicknacks").[33] To mark the success of Las Casas' theories in winning over the peaceful conversion of the Q'eqchi', the province was renamed *Vera-paz*, or True Peace, as it remains known today – a memorial of the territorialization of Lascasian hegemony.

In the Lascasian literature, many have debated whether this experiment was a genuine success. For his part, Las Casas saw the Verapaz as a vindication of his theories. In a letter written to the King on December 15, 1540, he explains his triumph: "Your Majesty mandated that I, with delegates of my order, would proceed with the pacification of the warlike provinces, and to put those of this region at the service and subjection of Your Majesty."[34] Las Casas aligns, quite rightly, the success of his approach with subjection, the production of new subjects, both in terms of their relation of subalternity to the Crown, and spatially ("those of this region"). Among contemporary writers there are those who argue that the ultimate failure of Las Casas' hopes for Tezulutlán is not an indictment of his thought:

> Had Las Casas met with more sympathy from Bishop Marroquín and other officials for his great plans in 1545, the bloody wars of the later period for the subjugation of the northern territories might not have been fought.... Las Casas's plans failed, *not because his ideas were wrong*, but because of the difficult conditions under which they were implemented, and because of the inadequate personnel and resources.... Today, after the collapse of colonialism, the efforts of Las Casas to secure the Christianization of the Indians by peaceful, voluntary means represent an early, shining, though incomplete application of the principles that should govern missionary work among non-Christian peoples.[35]

Our first reply to this sort of endorsement of Lascasian hegemony should be to note that the important question is not whether Las Casas' ideas were wrong. The problem with Lascasian hegemony is not its falsity, but its truth, its correctness, its sustaining power as *territorializing reason*. After all, from the perspective of the indigenous nations that were colonized, we are not "after the collapse of colonialism."

Moreover, the period of formal peace in the Verapaz was brief. Spanish colonists, often accompanied by Q'eqchi' collaborators, waged war against the *indios bravos* – mainly Manche Chol, Mopan, Lacandon, Itzaj', and other Mayas – who resisted the spatial violence of *reducción* that Las Casas strenuously advocated.[36] When Las Casas returned to Spain he insisted that the Crown establish a law – in effect, the first law of the Verapaz – that would make good on the promise made to the Maya leaders before their pacification and *reducción*: that the Spanish would not enter the Verapaz. At this point in the narrative, we should quote Remesal at length:

> In order to teach the natives to live according to the law...Las Casas addressed a memorandum to the Emperor in 1542 in which he made the following observations: He considered two measures essential; firstly, the establishment of *pueblos* [settled villages]. People should be made to live social lives in communities, otherwise a few isolated persons would pick up the new law and the rest would not, and in a state of dispersal they might not feel the obligation to live Christian lives. The sine qua non was a guarantee of absolute liberty, for if they did not enjoy freedom, they could not form a proper community and could not feel the obligation to be of service to one another. The saints say that because the first condition was absent, God did not give Abraham the law; no pueblo existed, only isolated tents.[37]

The law is a gift from God – given only to settled communities. Here a crucial element of Lascasian hegemony comes into view, and with it, Las Casas' justification of *reducción*. Colonization is to be secured through peaceful means, but not only in order to advance Christianity in a narrow sense. The Christian colonization of Tezulutlán and the subsequent enjoyment of "absolute liberty" can occur only by settling, through law as territorializing reason.

In 1538 Las Casas departed from his work on Tezulutlán for Spain, where he went directly to the Council of Indies to obtain the legal authority needed for his Order's work there. Remesal explains that Las Casas brought "the affairs of Tezulutlán" to their attention, and explained that his approach required a law to protect the Maya from the Spanish settlers:

> As everything he demanded was eminently reasonable and in accordance with law and justice, they granted him everything and...signed a series of legal documents.... One of them is of the utmost importance, for it prohibits the entry of Spaniards into the Indian territories.... "Fray

Bartolomé de Las Casas, OP, has informed us that he and...other breth-
ren of his Order have, by peaceful means and persuasion succeeded in
attracting to our service...the natives of the provinces and parts of the
kingdom of Guatemala, known as Tuzulutlán, and have induced certain
chiefs in these provinces to live with us on terms of peace....We hereby
order and command that in order to maintain this state of peace, no
person shall enter these territories."[38]

This is the first law *de Verapaz*, for it established the protection of the
Indians within sovereign territory by placing them within the sovereignty of the law – at the same moment that it constitutes this place
(*Verapaz*) as a space of true peace. Here we should recall Agamben's
argument in *Homo sacer* that "the paradox of sovereignty consists in the
fact that the sovereign is, at one and the same time, outside and inside
the juridical order."[39] As Agamben shows in his analysis of this paradox,
the sovereign exception is characterized by the fact that the sovereign
determines the conditions of possibility for law, stands outside the
boundary of the law, and may inaugurate or suspend the law.[40] The
same could be said of territory. Only a sovereign can declare that no
person shall enter the territory, can declare the boundary within territory
marking a zone of distinction. The act of declaring Tezulutlán under the
protection of the Crown – and this was, we should remember, an
essential condition of possibility for Lascasian hegemony there – reinau-
gurates Spanish sovereignty over the Maya. Remesal discerns the importance of this law in that "it prohibits the entry of Spaniards into *the
Indian territories*," yet the law only territorializes these the Verapaz
within Guatemala, within Spanish law. Still today within these spaces
there are no "Indian territories," no Maya states – only national territory
where the Maya are differentially included, that is, colonized. (The
recent Maya movements in the "Verapaz" – in both Guatemala and
Belize – have been condemned by the state as threats to national security.
Where, exactly, could we draw the Tuzulutlán/Verapaz distinction,
cartographically, today?)

Spacing Hegemony

Las Casas' experiment *de Verapaz* should compel us to space hegemony,
to ask after the spatial conditions of possibility for hegemony. In Gramsci's
notebooks, the concept of hegemony can be interpreted in three
different ways. First, hegemony can mean *leadership*. This meaning is
closest to the conventional use of hegemon (i.e., leading power), though

with Gramsci, hegemony in this sense implies class leadership mobilized through the establishment of moral or intellectual norms, i.e., leadership practiced through *persuasion*. Second, Gramsci uses hegemony to define a balance between coercion and consent that are employed to govern. Beyond a certain point, the costs of coercion – including the likelihood of provoking resistance – outweigh the benefits of the desired change. Hegemony names the point at which the constellation of forces that produces consent is achieved. And third, hegemony is sometimes used to characterize a condition in which one, or both, of the first two meanings is in effect. In this sense, hegemony is a condition in which a form of leadership is practiced that achieves consent without undue coercion.[41]

Gramsci's concept of hegemony is of great value for political thought, but it has its limitations. One of these concerns forms of geographical violence that facilitate the achievement of hegemony. What comprises hegemony if, as in the space of Verapaz, it is always already inscribed with a prior violence that did not obey the logic of hegemony? Is it not accurate to say that the Mayas of Tuzulutlán/Verapaz are the victims of more than five hundred years of violence and dispossession – and therefore there cannot be hegemony *there*? What could hegemony mean for the Maya, who have survived three conquests organized by territorializing colonial states?[42] Moreiras elaborates:

> The moment when the New World natives were led to admit that a Lascasian politics [including the relatively less violent practices of the British in southern Belize] were no doubt absolutely better than the practices of the cruel tyrants...that is the moment when Spanish [and British] *imperial* reason starts to become *hegemonic* reason: reasonable reason, post-terror reason. Reason after terror, but reason based upon terror: without the original terror primitive accumulation would never have gone into its second stage. Las Casas would never have converted a single native.[43]

The "natives" that Las Casas converted were Q'eqchi'-speaking Mayas of Tuzulutlán. *Before it was the Verapaz*. (This is the problem of geographical hegemony: we cannot write outside of or before the thematization of the world.) Before Christianity and its so-called "true peace" came to what is now the "Verapaz" of eastern Guatemala and southern Belize, these lands had other names in Maya languages. Las Casas tested his theory (the path to Christian colonization cannot be advanced by violence) on the Q'eqchi' Mayas.[44] His success at converting the Q'eqchi' made him one of the most loved fathers of the Catholic tradition. Was this, therefore, the decisive moment in the achievement of hegemony over the Q'eqchi'? No. The achievement of Lascasian hegemony – which facilitated the still-ongoing primitive accumulation of Maya lands[45] – already presupposed the more original violence and terror, in turn materialized through law and territory. The contrast with the experience, also in the same spaces, with the Mopan and Chol Mayas makes this clear: the promise of Las Casas' approach is nothing more than an offer not to suffer from a more abject terror. Lascasian hegemony continues to permeate the violence of imperial geography.

Lascasian Anthropology: Forerunner of Mayanism

> The quality of their minds is seen finally in superb artifacts, finely, beautifully fashioned, fashioned by hand. They are so skilled in the practical arts that their reputation should place them well ahead of the rest of the known world.... The practical things these people make are striking for their art and elegance, utensils that are charmingly done, feather work, lace work. Mind does this.
>
> Las Casas, *The only way*[46]

Las Casas is perhaps best known for the Valladolid debate, a great ceremonial contest held by the Council of the Indies against one of Spain's most learned theologians, Juan Ginés de Sepúlveda. Sepúlveda pointed to the Indians' lack of bravery, paganism, and practice of sacrifice to justify Spanish imperialism. Las Casas argued against violent colonization by asserting that all indigenous people were humans endowed with *rationality*, even where their conduct differed from European norms.[47] With this argument, Las Casas' humanism anticipated the formation of anthropology. The enormous text that he wrote to expand on his case in Valladolid, the *Apologética historica sumaria*, is nothing less than an encyclopedic elaboration of peoples from across the Americas, an attempt to prove their humanness through an analysis of indigenous cultures. (Though, strictly speaking, "indigenous culture" was not a concept available to or employed by Las Casas.)[48]

It is no accident that the man who is often considered the father of indigenous human rights is also the original American ethnographer. In this capacity, he wrote of the Maya. Las Casas was especially interested in the Maya because of his experiences in Verapaz and Chiapas, where he was the first Bishop (Las Casas returned to the Americas for the last time to take up this post, which gave him a means to support his Dominican brothers in the nearby Verapaz). Thus in chapter 123 of the *Apologética,* Las Casas narrates the discovery of a temple in "el reino de Yucatán," on the island of Cozumel, and then offers perhaps the first European description of the Maya gods Bacab, Echuac, Izona, and Cocolcán.[49] Though he interprets the roles of these gods in Christian terms – Echuac, for instance, is said to be the Holy Spirit – the tone of his text is ethnographic and empiricist: he describes what he has learned of Maya religion. And there is little doubt that his description is intended to provide for an appreciation of the gods. Whereas Bishop De Landa wrote about Maya gods as he attacked them, Las Casas wrote of Maya gods for the same reason he wrote about indigenous artifact production from across the hemisphere: to give an account of Maya reason within the reason of the West. By this Las Casas is distinguished as the preeminent forerunner of Mayanism.[50] With Las Casas the problematic of Mayanism becomes first encircled, first defined as a horizon: revealing the truth about the Maya in order to do justice by placing them within a European discourse about civilization. This is the becoming-space of Mayanism. It was already territorialized in the *Verapaz.*

Hardt and Negri observe that Las Casas' arguments against imperial violence extend from an anthropological theory of difference that presupposes a *European* sameness: "Las Casas cannot see beyond the Eurocentric view of the Americas, in which the highest generosity and charity would be bringing the Amerindians under the control and tutelage of the true religion and its culture. . . . Las Casas is really not that far from the Inquisition. He recognizes that humankind is one, but cannot see that it is also simultaneously many."[51] This limit of Las Casas' anthropological thinking was also, I stress, intensely geographical. His interpretation of the world and the unity of humanity was plainly derived from a Christian view that accounts for the world by way of God: the world came into existence the way God wanted it to; it reflects his essential goodness. Notwithstanding his appreciation for the reason reflected in the existence of Maya gods, Las Casas knew that there was only one God. He appreciated the unity of humanity – without radical otherness. The Mayanism of Las Casas was fundamentally a Christian undertaking, one that could acknowledge their gods because of the singularity of God's creation. His Mayanism was only the local result of a more general orientalism that embraced the world's diversity, but only in such a way that has prepared, in advance, a space, a slot, for every ethnos. A culture for every region and a region for every culture. Postcolonial critique aims to keep the play of worldliness alive, unsettling territorializing reason, to write a world that is everywhere one and many.

Maya Archaeology in the Archaeology of Mayanism

> The problem presented by the discovery of the early Maya settlements...
> is, briefly, this. In a region...overwhelmed by dense tropical jungle,
> entirely cut off from the Pacific and any possible Polynesian or Oriental
> influence by a lofty mountain range and wide tracks of forested coun-
> try...there exist groups of architectural remains, stone temples erected on
> lofty pyramids...[that] had been abandoned and practically forgotten at
> the time of the Spanish conquest.
>
> British Museum (1938: 7–8)

Las Casas is an exemplary figure who marks the becoming-space
of Mayanism. In his works, territorialization and the appreciation of
the Maya are joined in reason. Yet we cannot draw a direct line from the
Mayanism of the early 1500s to the present. To consider the conditions
that gave form to the Maya farm system discourse (see chapter 2), we
have to consider the ways that the Maya are reimagined, and Mayanism
reconfigured, as a consequence of the second discovery of the Maya in
the nineteenth century. Mayanism came into being after this second
discovery as an effect of posing, and trying to answer, the questions
that would shape Maya archaeology – especially two: who built the
temples found in Mesoamerica? And, what happened to their civilization? Mayanism originates in the search for clues to understanding the
ancient Maya civilization. Mayanist archaeology was fundamental to
this by producing a space for comparison of the ancient and modern
Maya, posing new questions about production methods and agricultural
capacities, and discerning the material nature of civilization. Agriculture
was seen as one of the strongest bridges between the ancient and contemporary Maya. Consider Gallenkamp's claim from 1976:

> Despite more than four centuries of economic exploitation, religious
> indoctrination, and social upheaval, the Maya have maintained a remark-
> able ethnic cohesion. Many tribes still adhere tenaciously to certain abo-
> riginal traditions – to the extent that ethnological studies of their culture
> have provided valuable insight into a wide spectrum of pre-Conquest
> practices.... Agricultural methods have remained essentially unchanged
> for thousands of years, with maize, beans, squash, chili peppers, and other
> venerable staples still being cultivated in *milpas*.[52]

One could find dozens of other passages like this one in the ethnographic
and archaeological texts of the Mayanist archive. Morley goes so far as to

argue that "the modern Maya method of raising maize is the same as it has been for the past three thousand years or more."[53] This view has been undermined by recent archaeological scholarship, but what is important here is not the veracity of such claims. The point is that such statements share important features, such as the comparisons between the ancient and modern Maya, and the allusion that the contemporary Mayas have lost what was truly great in ancient Maya civilization. This begins to explain how archaeology created the conditions for the Maya farm system discourse: the agriculture of the living Mayas attracted the attention of early Mayanists because it constitutes the ancient and contemporary Maya as a unity—as what transhistorically connects the Maya as ethnos to their land. At the same time, Maya agriculture is not simply one material link between the contemporary and ancient Mayas: "Maya agriculture…was the foundation of their civilization."[54] To show how Maya archaeology created the conditions for the Maya farm system discourse, I will briefly consider the work of three early Mayanists: first, the team of traveler-explorers, Stephens and Catherwood, who are often regarded as the discoverers of the ancient Maya world and the founders of Mayanism. I then turn to the work of Thomas Gann, a famous American Mayanist of the early twentieth century.[55]

Discovering American Civilization: Stephens and Catherwood

Until the middle of the nineteenth century, most Europeans either did not know or could not accept that massive ruins found throughout Mesoamerica were the material consequences of indigenous peoples. Theories abounded to explain the ruins as the work of Egyptians who crossed the Atlantic, or the lost tribes of Israel. Stephens and Catherwood's singular contribution was to combine a romantic portrayal of travel and discovery among the ruins of Mesoamerica with a sustained argument that the ruins resulted from an "American civilization." Stephens first advanced this thesis in his 1840 book, *Incidents of travel in Central America, Chiapas, and Yucatan*, a text filled with comparisons between the Egyptian, Greek, and Roman civilizations with the traces of Maya civilization. Stephens concluded that the ruined temples must have been "constructed by the races who occupied the country at the time of the invasion by the Spaniards, or of some not very distant progenitors."[56] (Note that the temples were identified *racially* from their second discovery by Europeans.)

In his two famous books Stephens musters several kinds of evidence for this argument. He frequently compares and contrasts the Egyptians and the Maya. In one passage Stephens describes the power of an idea that came to him as he was digging up a burial site in the ruins at Ticul:

> One idea presented itself to my mind with more force than it had ever possessed before, and that was the utter impossibility of ascribing these ruins to Egyptian builders. The magnificent tombs of the kings at Thebes rose up before me. It was on their tombs that the Egyptians lavished their skill, industry, and wealth, and no people, brought up in Egyptian schools, descended from Egyptians, or deriving their lessons from them, would ever have constructed in so conspicuous a place so rude a sepulcher.[57]

Stephens proves that the Maya are Americans, therefore, because they bury their dead with less care than the Egyptians. This is not his only evidence, of course. But his summoning of the Egyptian pyramids here is not capricious.

Stephens' frequent references to the Egyptians underscores the fact that Mayanist discourses were not simply created *de novo* in Mesoamerica. On the contrary: the "Maya world" of Mayanism was produced through the import of orientalist discourses to Mesoamerica. Archaeological practices traveled to "the Maya world" from other spaces, particularly Egypt, where they were executed by orientalists, colonial scientists, and amateurs. In the mid-nineteenth century British orientalists in Egypt and the "holy lands" were especially influential. Both Stephens and Catherwood learned and practiced their trade in Egypt. In 1822, Catherwood left England to travel through Rome, Greece, Libya, and Egypt, and in 1825 met Robert Hay in Alexandria; in 1829 Catherwood returned to Alexandria, where he was employed by Hay.[58] In 1837, three years after Catherwood returned from Egypt to England, Stephens published his *Incidents of Travel in Egypt, Arabia Petrea and the Holy Land*.[59] Catherwood and Stephens, drawn together by their common experience, decided to collaborate in exploratory travels through Central America. Together, they mapped, drew, and described several of the most important Maya ruins, including Copan, Tikal, and Palenque. Their two major co-publications were published in 1841 and 1843.[60] Most of these "views" present a landscape featuring ancient Maya ruins, covered partly by trees, and romantic figures – living Mayas – lying about in front of the ruins.

The texts would not have been so effective without Catherwood's evocative illustrations, which by representing the Maya help to produce them as ethnos. By portraying signs of ancient and living Maya, ruins and peasants, in the same scene, Catherwood demonstrates the proof of

Stephens' assertion that the temples were built by the ancestors of the living Maya. Yet even as Catherwood stresses the common, transhistorical Maya ethnos, he emphasizes the difference between the living and ancient Maya in often unsubtle ways. For instance, the living Maya in his works strike the most relaxed poses, as if unaware of the decaying, ancient ruins that frame them. Catherwood shows the indigeneity of the ruins by allowing us to see at once the ancient and living Maya together, but he craftily frames this relation as one that is separated by the division between the nature of the contemporary, *natural* Maya and the ancient, *civilized* Maya. His works defined the Maya as indigenous by locating them within a world-historical narrative of civilizations. Crucially, the Maya figures within his paintings are portrayed as though unaware of these divisions and relations. They do not read the complex signs on the ruins; they do not enter the temple passages; they cannot see into the space from where our gaze is directed towards them. These two bifurcations unfold simultaneously: Catherwood's staging of apparently innocent and casual scenes produces the Maya, ancient and contemporary, as well as the difference between us and them.

Consider figure 3.2, a typical Catherwood's painting, this one of the Casa del Gobernador at Uxmal. The scene juxtaposes living Mayas and an ancient Maya ruin. The presence of trees, fallen rocks, and earth emphasize that the great building has been ruined and that nature has overtaken this once-civilized space. This shift, from civilization to primitiveness, is emphasized by the figure of the snake. In the scene we find five figures, living Mayas, around a dead snake. A Maya woman and two children marvel at it, their faces filled with expressions of fear. But while the group sees only the snake, Catherwood invites us – the colonial viewer gazing in upon the other – to see that these figures, the five living Mayas and the snake, are themselves looked upon: by a snake figure, carved onto the ruin, shown above them and to the left. We see, therefore, that the ancient Maya, like their living ancestors, lived *here*, amid wild nature. And we know that they shared, like the figures in the painting, both awe and mastery over the snake. But whereas the living Maya have fundamentally a material relationship with snakes, the ancients had a symbolic one, manifest in the building's artful design; hence, the greatness of the ancients: they were great enough to choose to be able to represent nature. Catherwood's illustrations posit a relationship between the ancient and modern Maya through their co-presence in space. In this way his illustrations provide visual evidence for Stephens' argument that the ancient Maya built the temples – for the existence of the Maya qua ethnos.

Figure 3.2 Catherwood's painting of the Casa del Gobernador at Uxmal, 1844

Source: Catherwood, F. "Archway, casa del Gobernador, Uxmal." Plate X of *Views of ancient monuments in Central American, Chiapas, and Yucatan*, reprinted in Catherwood, F. 2000, p. 72.

Even as he establishes the common identity of the ancient and contemporary peoples, Catherwood underscores the decline of Maya civilization.[61] Catherwood's staging of the scene in figure 3.2 is an exemplar of Christian metaphysics, a romantic parable about sin and the fall. This Mayanist trope is described by Wilk:

> Archaeologists working in the Maya region ... often tell a little morality tale about how once upon a time the ancient Maya lived in harmony with

nature. But instead of staying happily in the forest, they tried to rise above nature by building cities and civilization, only to be destroyed through their own hubris, when they overexploited and destroyed nature.[62]

After the publication of Stephens' and Catherwood's popular books, and with the growing acceptance of their "American ruins" thesis, the region's Maya ruins came under greater scrutiny. Southern Belize soon became central to the map of the Maya world.

Thomas Gann's Explorations and Adventures

Thomas Gann, arguably the most popular Mayanist of his day, was not a trained archaeologist. An author of popular books describing explorations and discoveries, Gann led the early excavation of the two most important archaeological sites in southern Belize, Lubaantún and Pusilhá. The colonial state became aware that ancient sites were located in southern British Honduras at the end of the nineteenth century but did little besides confirm their existence and pass a law to prevent theft from the sites.[63] In 1903 Gann, then a medical officer in British Honduras, was charged to explore southern Belize. Gann's archaeological work was an extension of his role as a scientist employed by the colonial state.[64] His later work relied on the British Museum and the British Honduras colonial state (see figure 3.3). There is a long history of such coordination between archaeology and European colonial states, particularly the British, French, and US. Colonial power solicited Mayanist knowledge and facilitated the access and control of archaeological sites.[65] Gann wrote a monograph on Lubaantún that was republished in London by two different anthropology journals. In 1924 and 1925 Gann returned to work at Lubaantún, research that led to the publication of his third book, *Mystery cities: Exploration and adventure in Lubaantún*.[66] *Mystery cities* is a breezy text, written by a colonial medical officer, combining elements of archaeology, travel writing, and ethnography.[67] The narrative is driven by Gann's hurried explorations: Gann is anxious to keep the pace moving, and yet also to demonstrate that his excursions are not mere travel, but are producing real knowledge about the Maya. So although it cannot be denied that *Mystery cities* is a product of ethnographic description and archaeological study (as characterizes most Mayanist work), Gann's shamelessly amateur tone and racist asides make the text something of an embarrassment for the serious Mayanist. Notwithstanding these faults or the criticisms Gann's

Figure 3.3 The British Museum Expedition Team, 1928

Source: Gann, Thomas, 1928. *Discoveries and adventures in Central America*, p. 102. The alliances between the colonial state, the "Maya explorers", and the museums backing Mayanist research were often formal and explicit. This photo, taken from one of Gann's popular accounts of his explorations, we find aligned (from left to right): Mr. Clive-Smith, a representative of the British Museum; Sir John Borden, the Governor of British Honduras; Thomas Gann, the world's best-known Mayanist at the time; and Captain Gruning of the British military. Clive-Smith and Gruning "carried a perfect arsenal of weapons" to protect the expedition. These four led the 1927–28 expedition to the ruins of Pusilha in southern Belize. Their trip was sponsored by the British Museum—which still holds many of the artifacts they obtained.

work at Lubaantún faced from later Mayanists, there is no doubt that Gann was considered a credentialed Mayanist in his time. He published seven books on the Maya, and co-authored his *History of the Maya* (1931) with the doyen of twentieth-century Mayanism, J. E. S. Thompson.[68] Most of Gann's works aimed at the popular British audience, as he exploited the nostalgic demand for the unabashed colonial exploration books of the Victorian era.[69]

Gann's texts iterated the Mayanist theme that the contemporary Maya were rooted essentially in the *milpa*, tied to their ancestors by agricultural practices. He rarely misses an opportunity to mention ancient spirits when he passes a *milpa*. In *Mystery cities* Gann frequently complains that he found it difficult to motivate the living Q'eqchi' residents of San Pedro Columbia, the community that is home to Lubaantún, to work at the ruins for wages. The problem, as Gann explains, is that "for nearly 2,000 years the Indian has relied upon his plantation to supply him with food, drink, clothes, and luxuries, till its cultivation has become a sort of sacred ritual."[70] The Mayas, back to the time of the ancients, have known no real labor. All their life is a party: "The communal work on the *milpa* is really a picnic, in which all the men of the village join, their corn cakes, beans, and *sacha*... being brought out to them at mid-day by the women and children."[71] Passages like this, where Gann suggests a romantic view of Maya agrarian life, are always qualified. As Gann explains, this bucolic way of life, where real labor is unknown and families pray together in the *milpa*, is doomed by the mixing of races and the continuing collapse of Maya civilization:

> Unfortunately, the Maya Quiche, as well as the northern Maya, ... are gradually losing their own *milpa*-making habit, and so, cut off from the land which they love, from the ancient agricultural tradition and age-old mode of life, they are rapidly approaching the last stage in that slow process of degeneration which began nearly 1,300 years ago, when they left their Old Empire cities to migrate into Yucatan, and which can only end – and that before many years have elapsed – in complete extinction.[72]

Gann's text here condenses one of the central themes of Mayanism. He appreciates the Maya as rural people, who know no real labor; he admires their ability to cultivate connections to their ancestors through cultural and agricultural practices; and yet he must reveal that they are doomed to "complete extinction" with the decline of their organic ties to the land and the Maya farm system. The extinction of the Mayas has been unfolding through a "slow process of degeneration which began nearly 1,300 years ago, when they left their Old Empire cities." If we accept that their fall from civilization exists; if we allow "that process" to obtain an ontological quality and therefore a teleology: then, indeed, the complete extinction of the Maya – all of them, since they are a cultural unity – does seem inevitable.

Gann, like Mayanists before and after him, is interested in articulating the ancient and the modern Maya, and like many others he does so in the *milpa*. Agriculture is taken to be the key to understanding the Maya fall from civilization. Consider his comparison of the ancient and contemporary uses of space, and quality of life, which he sketches in his discussion about what Lubaantún must have been:

> When these ruins [at Lubaantún] were inhabited the population must have been a very large one; what is now an unbroken stretch of sunless, airless, almost uninhabited bush was then a smiling, open country, covered with maize, intensively cultivated, the scene of a busy agricultural civilisation, open to the sun and breezes from the sea, and free from malaria and the innumerable insect pests which now make life a misery to the archaeologist or explorer traveling through it. In fact, what was during the old Maya Empire a pleasant and salubrious country is now just the reverse.[73]

We can find here clear parallels between Gann's text and Catherwood's portrayal of the Casa del Gobernador at Uxmal: what was once "a smiling, open country, covered with maize" has suffered a fall: all that was "pleasant and salubrious" is now "just the reverse." What was once great is now uncivilized.

Let us review the two main lessons of Mayanism. The first is that the Maya exist, that they are a race, and that their racial essence was formed over thousands of years in the Maya area. In different ways, all of these texts point to the essential relationship between the Maya of different times and spaces. They do so to produce the Maya qua ethnos. Consider figure 3.4: Producing the "Maya." When Gann and his team excavated discovered Lubaantún, they uncovered a small jade carving of a person's face. In *Mystery cities* Gann places a photo of the carving alongside of a photograph of a "head of little Maya girl of 6 years," as the caption explains. This figure appears without other reference at a point in the text where Gann compares and contrasts the "Lubaantun Indians," as he calls the local inhabitants of the nearby village of San Pedro Columbia, with the ancient Maya. Placing the two heads together the Mayanist sets up the Maya qua ethnos, within a space of comparison that is never defined. It is of course a European identification of the Maya, one that

MAYA CHILD.   JADE CARVING. 1,500 YEARS OLD   HEAD OF LITTLE MAYA GIRL OF 6 YEARS

Figure 3.4 Producing the "Maya," 1925

Source: Gann, Thomas. 1925. *Mystery cities: exploration and adventure in Lubaantun*, p. 162. Images that place the ancient and living Maya alongside one another proliferate in early 20th century Mayanist texts. This image, from one of Thomas Gann's celebrations of "exploration and adventure" in Belize, is one of many that produce the Maya through the bifurcation of the category "Maya" into "ancient" and "modern" and the re-presentation of each.

appreciates the existence of an other ethnos, one that we can know objectively. Literally so, through objects like this one, 1,500 years old.

The second lesson is that after a brief, glorious, "pleasant and salubrious" period, the race went into a permanent decline. Gann's texts speak frequently about this continuing decline. Writing again about his troubles relating with the local laborers, he explains: "The Kekchi, in fact, has descended the ladder of degeneration several steps lower than his northern brother, and has got uncommonly close to the bottom."[74] Gann's texts are replete with teleological expressions like "the ladder of degeneration." A few pages earlier in the text we find that "the Indians of the neighborhood are a poor, feeble, anaemic, degenerate race, representing the last step of the once great Maya people along the road to extinction."[75] This particular racist claim is followed by an enormous question, one that offers a clue to reading the Mayanist narrative and its relation to discourse on the Maya farm system:

> Living, as they do, amongst the tombs of their ancestors, whose remains for fifty generations surround them on all sides, over whose graves they make their corn plantations year after year, amongst the ruins of whose stupendous cities they constantly wander, is it to be wondered at that they are a ghost-haunted, spirit-obsessed people, living more in the past than in the present, concerned more with spirits of the dead than with the living, and that they dream dreams, see visions, and hear sounds not of this world – that the dense forest is peopled for them rather with the shades of the countless dead, who occupied it for nearly 2,000 years, than with the small, and rapidly vanishing, mournful little remnant, who seem to be only awaiting patiently the fulfillment of the time when the last Maya shall have joined his forefathers, and complete silence settled over the dense veil of immemorial bush which shrouds the temples and palaces, the cities and the monoliths, where once flourished a vast population, and the highest aboriginal civilisation ever known in the New World?[76]

Confronted by such brutal essentialism, we should pause to recognize the complicity of his concluding, respectful remark, that the Maya are the stewards of "the highest aboriginal civilisation ever known in the New World." The Mayanist appreciates and respects the Maya, and it is *on this basis* that he can ask: "is it to be wondered at that they are . . . living more in the past than in the present?" Like his iterated gesture to the inevitable decline of the Maya, Gann's question indicates a narrative that structures Mayanism: the Maya are living out of time.

We saw earlier that two questions have driven Mayanism from the start. The first was establishing Maya as an *American* civilization. No doubt the increasingly imperial character of the late nineteenth-century US raised the pertinence of this question. But the longest-standing question concerns the Maya collapse. We are still faced by this question. At the heart of the debate, the way the question is framed, lies the Maya farm system. This system, this organic tie between the ancient and the living, has long been looked to for some explanation of the failure of the ancient Maya to sustain their civilization. Consider Lundell's proposal to reconstruct ancient Maya civilization on the basis of his study of the Maya farm system:

> Having indicated briefly the essential points of the present [Maya farm] system, I should like to apply the data in making a hypothetical

reconstruction of the life and agricultural system of the ancient Maya. In so doing I shall assume that the stage of civilization to which the Maya have dropped back is not only a stage through which they passed in the development of their higher culture, but that it is a stage which was never outgrown at the periphery of the Maya region, and was probably not outgrown even near the center of the region in backwater districts. The present level is one which has presumably always been familiar to the Maya, one to which they could revert even as it was one from which they could advance. To understand it is therefore essential to the higher levels of the Maya culture.[77]

This clarifies the debate about the Mayas' farming, in ancient and contemporary forms, and the question of the relation between the two. But we cannot say that this question *determined* the form of the Maya farm system discourse in advance. It is rather that the archaeological questioning of the decline of the Maya, in a distinctly orientalist mode, created the conditions for the Maya farm system. It did so by calling for some unity to be named as a system: precisely a system that could be seen, named, grasped, and analyzed as a complex unity that stretched from the ancient Maya to the living Maya; that in its essence and its long arc of decline could be read the rise and fall of Maya civilization. The elements of the Maya farm system discourse – the existence of the Maya culture; the link to agriculture; the destruction of the forest; the essential tendency to abandon and scatter – all these were made available by Mayanism. The question was if and how it would be assembled in speech. And to what effect.

The Maya in Asturias's *Hombres de maíz*[78]

Miguel Ángel Asturias (1899–1974) became the first Latin American prose writer to win the Nobel Prize in literature, largely for his novel *Hombres de maíz* (1949). This enormously complex work, surely one of the greatest novels of the twentieth century, has been credited for inspiring magic realism. Drawing inspiration from the *Popol vuh* and ethnographic studies of the Maya, the novel's drama, woven through six separate but interdependent stories, is motivated by Maya characters who emplot and partly narrate a morality tale about distinct national identities, territory, and the deepening of capitalist social relations.[79] The book's enigmatic title reflects a crucial distinction made in the novel between two subject-positions: the *maicero* (or "maizegrower") and the *hombre de maíz* ("man of maize"). For the *maicero*, maize is a commodity produced as a means to an end; for the *hombre de maíz*, maize is at once means *and* end, part of an unbroken circle of life and culture in which the people of maize constitute themselves through the cultivation and consumption of maize. The *maicero* does not extract all the value that is generated from the destruction of the forest. The *hombre de maíz* integrates himself and his culture *with* the forest: his social relations are not premised upon alienation, they are not essentially capitalist:

> The matapolo is bad, but the maizegrower is worse. The matapolo takes years to dry a tree up. The maizegrower sets fire to the brush and does for the timber in a matter of hours. And what timber. The most priceless of woods. What guerrillas do to men in time of war the maizegrower does to the trees. Smoke, flames, ashes. Different if it was just to eat. It's to make money. Different, too, if it was on their own account, but they go halves with the boss, and sometimes not even halves. The maize impoverishes the earth and makes no one rich. Neither the boss nor the men. Sown to be eaten it is the sacred sustenance of the men who were made of maize. Sown to make money it means famine for the men who were made of maize.... The earth will become exhausted and the planter will take his little seeds off somewhere else, until he too begins to waste away like a discolored seed fallen in the midst of fertile lands ripe for planting, lands that could make him a rich man instead of a nobody who wanders around ruining the earth everywhere he goes, always poor and finally losing all pleasure in the good things he could have had.[80]

That is, the *maicero* is a peasant who does not own the land that he uses, and he does not farm the land that he is from: he rents land from a landlord and moves frequently.[81] The *hombre de maíz* is a precapitalist, indigenous Maya farmer who lives transhistorically on traditional lands. If the itinerant *maicero* can only wreck environmental destruction and the social relations of exploitation in his wake, the *hombre de maíz* is bound to the land, the ground of national territory. On the basis of these distinctions, *Hombres de maíz* criticizes capitalist agriculture: it is the compunction to accrue capital that compels the capitalist and pushes the *maicero* to clear the forests, plant maize to sell it, and displace the *hombres de maíz*. The moral of the novel is that this encroachment of capitalist social relations into Maya territory threatens the nation.

Although Asturias was born and raised in Guatemala, where *Hombres de maíz* is set, it was in Europe that he became inspired to write his great Mayan novel. In Paris, Asturias studied with Georges Raynaud (then one of Europe's leading Mayanists) and assisted in translating the *Popol vuh.*[82] He also traveled to London's British Museum, where he was enthralled by its Maya collection.[83] There he found, in the words of Luis Harss and Barbara Dohmann, "scarecrows out of his own past":

> They were a mute reminder that although time and distance had effaced the tattered splendors of the old Indian civilization, its vision of the world and its modes of thought were not entirely gone. He had caught glimpses of them at home, dormant, fossilized in an inscrutable population reduced to misery and despair. But their signs could still be read.[84]

Asturias's apprehension of the museum's Maya artifacts – mute, effaced, tattered, old, dormant, fossilized, inscrutable, in misery and despair, to be appropriated and made intelligible by the trained specialist ("their signs could still be read") – inspired him to try to relate "the old Indian civilization" and "its modes of thought" with *Guatemala*, his "home."[85] Although he had of course known of the Maya before Europe, he did not know them as an appreciative Mayanist. On the contrary: in 1923, just preceding his departure for Europe, Asturias argues for a eugenic approach to Guatemala's "Indian problem" in a law thesis called *El problema social del indio*: "the Indian represents a past civilization and the mestizo, or *ladino* as we call him, a future civilization."[86] To recalibrate the Maya with modernity, Asturias proposed that the "Indian character" should be genetically overwhelmed through mass immigration from Europe and state-sponsored *mestizaje*: "to solve the present problem of the Indian...we need to transfuse new blood into his veins."[87] His appeal to eugenics is grounded upon a discourse about national space. In a move that he will repeat throughout his work, Asturias seeks to territorialize the Maya qua Guatemalans; to facilitate Guatemala's unification and progress, Asturias argues, "the 'feeling of the land,' as Nietzsche calls it, must serve us as intuition. With its assistance we will study *our piece of the globe* (its territory) and *the people that inhabit it* (its population)." Only such study could "provide us with the opportunity to make a racially, linguistically, and economically homogenous nation of Guatemala."[88]

Having thus argued that saving the Maya required eliminating them, he returned only a decade later from Paris and London, inspired to the project of retelling Maya stories. This shift from eugenics to Mayanism was not propelled by voices of Maya people in Guatemala; Asturias's aesthetic and political conversion occurred in Europe. Again, the site of his transformation are the British Museum, and his work with Raynaud – in Mayanism – where we find the bridge from his brash positivism to his incipient magical realism. From eugenicist to indigenist, Asturias went from treating Mayan speech as a sign of intellectual weakness to admiring its secrets and imitating its rhythm to produce a new literary voice. Asturias advanced Mayanism by drawing upon and reproducing its truth – *poetically.*

Asturias's rearticulations of space, territory, and race coalesce in the *milpa*, where he locates Maya resistance in what he understands to be ancient Maya farming practices. He thereby reiterates one of the central themes of Mayanism, i.e., the *milpa* as the site of authentic, precapitalist Maya culture. At the beginning of the novel, the narrator explains that Gaspar Ilóm takes up the defense of Maya lands because the *maiceros* are destroying the forest, impoverishing the earth and bringing famine to the *hombres de maíz*. Driving these processes are the expansion of capitalist social relations: the *maiceros* act as they do not on their own account, but in order to make money on someone else's behalf. In a novel that is remarkable for its magical metaphors and surrealistic settings, it is striking to trace the way in which this conflict becomes a kind of morality tale, reiterated in straightforward fashion four more times. Musús explains to Colonel Godoy, "these maizegrowers are dull-witted folk, colonel. They see danger coming, but they don't get out of the way. The maizegrower in the cold lands dies poor or dies murdered. The land is punishing them through the hands of the Indians. Why sow where the harvest is bad? If they're maizegrowers, why don't they go down to the coastlands...?"[89] In the midst of the war against Gaspar Ilóm's men, this Ladino soldier maps the correct racial geography of Guatemala: Ladinos do not belong in the "cold lands," the highlands – that is, Maya territory. Later, Benito Ramos repeats the lesson: "You should have seen what this land was like when [the Indians] were cultivating it rationally.... Maize should be planted as they used to plant it, as they still do, to give the family its grub, and not for business.... [T]he Indians used to... [do] things in a small way, if you like, but they had all they needed, they weren't greedy like us because now...*greed has become a way of life to us.*"[90] Compare Benito's claim that all was well when the Indians "were cultivating it *rationally*" with Asturias's stated source of inspiration, explained at the Nobel award banquet, where he claims his work was an effect of "cataclysms which engendered a geography of madness, terrifying traumas, such as the Conquest:... Continents submerged in the sea, races castrated as they surged to independence, and the fragmentation of the New World."[91] The land-Maya articulation was a rational system; capitalism has broken this tie; today the geography of the Americas is mad.

The same moral is repeated twice by Maya characters. First the Old man with Black hands explains that producing maize as a commodity brings destruction ("maize costs the sacrifice of the earth.... [W]hat they're doing now is even more uncivilized, growing maize to sell it");[92] again, in the novel's final pages, the Curer-Deer of Seven-Fires laments the "maizegrowers who sow maize in order to profit from their harvests. Just as though men made their women pregnant to sell the flesh of their children, to trade the life of their flesh, with the blood of their blood, that's what the maizegrowers are like who sow, not to sustain themselves and support their families, but covetously, to make rich men of themselves."[93]

Four characters, five passages, but one set of guiding truths: maize should not be commodified; the commodification of maize will impoverish the ladinos and eliminate the Maya – thereby will unravel the tie between the nation and its territory. Asturias thus draws a new map of Guatemala, but one that remains governed by the becoming-spaces of race. By framing the capitalism-territory-Maya relation in this way, he impinges upon and limits the possibilities of other worlds that his narratives might open up. The figure of the Maya precapitalist farmer, his place in the world, his land – all these are constructed as interdependent and interlocking parts of a whole. Spatially, it is within the *milpa, and only there*, that they join, that the system works. The argument becomes a romantic Mayanism. But while *Hombres de maíz* is true to the archaeology of Mayanism, it is less true to the complexities of Guatemalan social relations. The driving dynamic of *Hombres de maíz* is established as the conflict between the indigenous *hombres de maíz* and the ladino *maiceros*, driven by capital and the landlessness of the *maiceros*. Had Asturias rigorously shaped the novel around Guatemalan social relations, however, this dynamic would have to recognize that the very *maiceros* who are compelled to "waste away...in the midst of fertile lands" given over to plantations of sugar cane, banana, cacao, coffee, and wheat were often *land-poor Maya peasants*. In other words, the *maiceros* are *hombres de maíz* – Mayas who have already been dispossessed of their lands.[94] *Hombres de maíz* thus frames the relations between race, space, and class in ways that cannot treat the complexity of Guatemala's social classes, their inflections through racial discourse, and their historical-geographical conflicts.

Asturias's own life experiences contribute to the novel's conditions of possibility, and complicate the moral of *Hombres de maíz*. To the extent that Asturias's critique of capitalism encourages a turn toward the maize *milpa*, it implies equally a turn away from the wheat-selling village store. For just as maize is treated as the source of life for the men of maize, wheat registers negatively. The *maiceros* are said to become poor by leaving their farms to go to work on banana and coffee plantations in western Guatemala "in rich soil spattered with blood, and wheatfields ablaze beyond."[95] (Asturias, like Diego Rivera, tropes wheat as a genetic intrusion into American maize-fields, a symbol of European imperialism.)[96] Remember that upon return from Europe, Asturias desired to tell Maya stories, to translate their speech in a more universal mode. Yet as an urbane, bourgeois ladino, his opportunities to interact with the Maya were limited by social and geographic distances. How then did he collect the stories that comprised *Hombres de maíz*? The answer hinges on wheat. Asturias's studies of the Maya were facilitated by his father's trading business. Maya men came to his store from rural villages, to sell maize and beans, to buy sugar and flour. They often traveled to the town to trade, and would sometimes stay overnight in the Asturias compound before making the journey home:

> Asturias's father had become an *importer of sugar and flour, which he sold to the people who inhabited the surrounding countryside*. He held a constant open house to accommodate his clients, and the gatherings that took place in the courtyard at nightfall, under the trees, were an endless source of wonder and information for the young Asturias.... [Asturias recalls that "t]he buyers came in on their carts, or driving their mule teams. They arrived in the morning or the afternoon, did their marketing, then packed their loads to be ready to leave the next morning. They spent the night in the courtyard.... I heard them talking every night, telling their stories."[97]

There, in the courtyard of the compound, he chatted and listened to the stories of these Maya farmers engaged in capitalist social relations. The condition of possibility for *Hombres de maíz* is the marketplace where corn becomes a commodity. The son of a man who made a living selling *wheat flour* would write a novel celebrating his father's customers – as precapitalist men of *maize*.

Hemmed in by a racialized map of the world, Asturias's narrative cannot offer any kind of articulation between competing economies, between wheat and corn. While corn denotes a pure, indigenous way of life, wheat registers as the sign of Euro-ladino impurity. Race intrudes upon his social model, setting up pure spaces that cannot accommodate capitalist Mayas or precapitalist, ladino, *milpa* peasants. Two elements of the discourse about the commercialization of maize further mark these limits. First, Asturias emphasizes that poor ladino *maiceros* are stuck within an oppressive capitalist system, while the Maya live *outside* of (and under attack from) capitalism. Yet the Maya are not outside of capitalism. Indeed, the liberal reforms that provide the historical backdrop for both his life and his literary efforts were precisely intended to (and did) deepen capitalist social relations in Maya communities. Unlike Marx, then, Asturias locates resistance to capitalism strictly *outside* of itself – not within capitalism as a mode of production, but *before it*, in what Marx would have called "primitive communism," a stage from which the Maya had exited long before. Second, Asturias treats the relation between the Maya and the land as both *essential* and *ahistorical*. As he would later write in the introduction to a coffee-table book on Mayan art: "It was these civilizing gods that taught the inhabitants of the plateaus of Guatemala . . . *to cultivate maize, exactly as it is sewn and reaped today.*"[98] But Maya livelihoods are not ahistorical. Maya farmers have long participated in the roiling and exploitative social relations we call "capitalism": they often rent land, and sell their labor; they also sell corn and buy commodities like wheat and sugar, when they can, as Asturias would have known from his experience in the courtyard. What is at stake here is a way of conceptualizing the subsistence practices of rural Maya communities: *Hombres de maíz* rigorously, too rigorously, separates historical categories – ancient and modern, subsistence and capitalist – in terms that ultimately derive not from Maya but rather Mayanist narratives. These temporal distinctions place the Maya, again, outside of capitalism and inside of Guatemala. Since the historical terrain of their struggle for justice must emerge from the precapitalist *milpa*, the authentic spaces of Maya resistance are territorialized in advance – as potentially national land.

Asturias's critique of capitalism's entry into the *milpa* relations therefore naturalizes one particular race-space articulation: Mayas are maize, the marginal-yet-central people through which he imagines a rearticulated Guatemala. Who inhabit, of course, the naturalized Indian home at the very center of Mayanist discourse: the *milpa*. The becoming-spaces for Guatemala's races are thus cleared. It remains only for each group to be placed within its territory. Like a powerful god, the hand of Asturias pretends to intervene in this world from outside of history. Expressing a desire born in the heart of Europe, articulating words accumulated where wheat and maize were traded, a magical tale of resistance opens a space in history only to place the Maya within their *milpa* on the margins of Guatemala.

The Maya in the British Museum

If we were to insist that *Hombres de maíz* was the progenitor to Latin American magical realism, and likewise take Asturias at his word to believe that he "discovered" the Maya, his roots in Guatemala, and the problem of writing of *that* reality in the British Museum, then we cannot but conclude that the Maya *of* the British Museum can take credit for magical realism. This would be a delightfully subalternist conclusion were it not also ridiculous, for who exactly should get credit for thoughts inspired by the British Museum? (Should we thank the museum's famous library for the first volume of *Capital*?) To whom or what can we attribute Asturias's inspiration? The museum? The Maya? No, we would have to say: the Maya *at* the British Museum. But what, or who, are the Maya at the museum? Or, perhaps we could ask, to abide by our theme of spacing: *where* are the Maya in the British Museum?

But why the British Museum? Not only because Asturias found himself there, no. The British Museum is one of the institutions most responsible for Mayanism. It was a pivot-point around which archaeologists with experience in Egypt, the "Near East," and elsewhere came to work in British Honduras, Mexico, and Guatemala. Alfred Maudslay, Thomas Gann, J. E. S Thompson: all received support from the British Museum, and in turn many of their "collections" are there today.[99] The Museum served, in Latour's words, as a center of calculation for the observation and analysis of the Maya.[100] And the British Museum is not simply one museum among many, or even one of the greats: it is indeed The Museum (as it is often called), the model of what a European museum that represented the best of the world could and should be.[101] It remains the model "world museum," one that "holds in trust for *the nation* and *the world* a collection of art and antiquities."[102]

Repository of the spoils of the Empire, the museum today claims ownership of the lintels of Yaxchilán, the Rosetta stone, and (most infamously) the marbles that Lord Elgin stole from the Parthenon.[103] Maya, Egyptian, Greek: by setting up their artifacts within a framework of geographically defined rooms, the museum constitutes a space of imperial comparison – in a dual sense. First: it is itself a product of British empire, an effect of the British colonization of the Americas, Africa, and the Orient. Without this history, fewer jewels would be kept in Bloomsbury today. Indeed, the museum was an active agent in colonization in many places, funding expeditions and providing a moral and scientific justification for exploration and the removal of "objects."

Second: the museum is itself a space for contemplation of empire and civilization – their nature, heights, and predisposition toward collapse. There are no more elaborately constructed, no more perfect sites in the world, within which to watch civilizations rise and fall as one strolls from room to room.

To return to our question: *where* are the Maya in the British Museum? Today, in the Mexican Gallery (Room 27), which can be reached by walking through North America (Room 26, sponsored by J. P. Morgan Chase). Some great nations have their own room in the British Museum (China, Korea, Greece, etc.) but all the native peoples of the Americas share Room 26 – except for the Olmec, Maya, and Aztec people, who share Room 27. The Museum's description of the Room 27 explains: "ancient Mexican civilizations left a dazzling legacy for modern museum visitors. Jade and stone sculptures, turquoise mosaics, painted books, hieroglyphic inscriptions and cast gold are a few of their surviving artistic achievements."[104] Defining the Maya's place within the room, we read: "The Maya (AD 300–900) inhabited sixty separate kingdoms and their highly sophisticated society produced a complex writing system and beautiful sculpture and ceramics.... The tour uses objects in the British Museum's collections to examine what we know about... their rise and fall, rulers and citizens, rituals and beliefs."[105]

Walk through the Americas room and you enter the Mexican Gallery, a rectangular space interrupted by angular displays filled with artifacts of jade, clay, gold, and stone. Once a cataloging room, the new gallery is designed to evoke both temples and tombs. As one archaeological review explains, "the aim has been to exploit and express the inherent drama and primeval potency of the exhibits... [and] create an atmosphere that consciously refers to original Mesoamerican architectural forms and materials."[106] The Maya are represented today by the highlights of the British Museum's Maya collection – a collection that once comprised its own room (the very one that inspired Asturias) with its own museum guide. The lintels of Yaxchilán are set laterally into a red wall (they faced downward *in situ* at Yaxchilán). In the glass case that displays the finest works of the Maya collection, there is a triptych of jade carvings (see figure 3.5). The center of these is the jade head featured in Gann's book as the illustration of Mayanness – our familiar jade head, disinterred at Lubaantún.[107]

Although Room 27 is the Mexican Gallery, in light of the Maya objects from the present-day territories of Belize and Guatemala, it could equally be called the Mesoamerica Room. But in fact the room represents neither simply Mexico, nor Mesoamerica, but precisely the becoming-space of Mexico qua national territory. For what distinguishes these three peoples (Olmec, Maya, and Aztec) is not their Mexicanness *per se*, since, after all (a) there are other indigenous groups in Mexico *not* included in Room 27; (b) the Maya exceed the territory of Mexico; and (c) at the time these civilizations were at their height, there was no "Mexico." What collects these three civilizations together and binds them in this room is their designation as foundations of modern Mexico. Notice the capricious significations mobilized by the jade head from Lubaantún: in Gann's text it marks the essence of the Maya as deterritorialized ethnos. Locked in a glass case in Room 27, it stands in for the achievement of a family of peoples collected, today, under the sign of "Mexico."

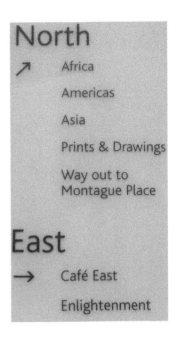

Taking these objects, texts, and spacings together, Room 27 not only reiterates the classic Mayanist question ("Why did this these civilizations collapse?") but, more fundamentally, presents to the pedestrian viewer an experience of geographical-ethnographic objectification that makes *that* question possible. And as an *empirical* question. But how? The viewer is in London, in the British Museum, presented with objects (mediated by texts) that represent Mexico in general, Maya-Mexico more specifically, Yaxchilán in particular (home of the lintels). And these are arranged within a temporal scheme structured by the development (as unfolding) of civilization. The object-texts in the museum both portray this other place and also stand in for it. As portrait they are delightful visual samples of people and places – different from us, far from here – from which we are separated by space and time. As proxy, they form one small part of a whole ensemble, as the cultural-object representative of a people within the broad panoply of the history of human achievement and civilization. Although some peoples and places receive more attention than others, by definition all are accommodated; there are no peoples left out. Like Las Casas' theology, the museum recruits them all.[108]

(A)

(B)

Figure 3.5   The Maya in the British Museum. (A) A view of the Maya area in the Mexico Gallery, with one of the lintels of Yachilán on the wall. (B) Inside the glass case featuring the treasures of the Maya are these pieces of jade

Source: J. Wainwright.

At the moment when the pedestrian viewer sees an object within Room 27, we may ask: isn't the museum-goer an agent, a critical thinker, able to read and think about what is presented? Certainly. But even before the viewer visually takes in the object, there is already the tour guide, the views of the expert directing them via text, the work of the architect and designer arranging the objects in space, etc. The experience of seeing the object is mediated by the texts and spaces that frame it. But this is only the immediate and relatively concrete frame for relating to the object. Because the specifically "Maya" object, for instance, is defined geographically – again, it represents Mexico, "the Maya world," and the Americas – it is taken as a world-orienting compass. The object, whatever else it may represent, stands in for this other space. The jade head, again, represents premodern Mexico. This effect of orientation is a product of a relation between the viewer and – not the object, which is simply *there* – but the wider world pointed to by the object. The viewer is spatially interpellated within this worldliness, as the fact of the jade head's being *there* calls the viewer into a relation with Mexico/the Maya. It is *this* relation that we sense when we feel the museum's orderly and totalizing density: literally the weight of the world. That this "individual experience" (of being interpellated by a place/people) via an object is not possible just anywhere reveals that the effect is produced by the museum. The museum is organized in a performative, geographical manner. Different people-places are afforded different spaces, but again, all are recruited. The relationship between the people/place signaled by the jade head and the viewer is only possible because it is slotted to be *there*, framed by the signs that point out where we are: Korea right, Japan left, and in Room 27, Mexico. We discover, in the jade head we see there, *Mexico* – and yet, again, this jade head was taken from "Belize" (or Lubaantún, or San Pedro Columbia – where was it found?). What we are encouraged to see in the jade head here is a fragment of the civilizations that *preceded* Mexico, the becoming-spaces of Mexico. Precisely as Miguel Asturias discovered Guatemala through the Maya in the same room.

It is this geographical staging of the world that is the essence of the British Museum – and, if Heidegger is correct, also the essence of modernity.[109] The organization of the museum to represent the world as such could not be thought or imagined in Europe before the Enlightenment; it reflects Europe's self-recognition as a geographical people/idea.[110] Over time the British Museum has come to be organized geographically, not thematically, to represent the world in microcosms of art and artifact, where the goods of worldly techné are arranged to behold. Where derives the desire for this view, this power? Is it British? Would that it were: if so, we would not find so many imitations around the world. When the Mexican Gallery was opened in 1994, the Mexican Ambassador "expressed the sentiment that his country was proud to have a gallery dedicated to its culture alongside the other great cultures represented in the British Museum."[111] (Yet Room 27 makes no mention of living descendents of the Maya in Mexico, the internal heterogeneity of the various "pre-Mexican" civilizations, the support of the Government of Mexico in the production of the Mexican Gallery, the origins of the British Museum's collections, nor the colonization of Mexico by the Spanish or Belize by the British.)

No, the desire to frame the world in this way has spread in and through the concept of civilization. The arrangement of the world, these techné of every people-place, reflects nothing except the valences and the matter of Western civilization's theory of civilization: (a) all cultures have value via their technical-artifactual production; (b) they may be arranged and compared *empirically*; (c) doing so reveals progress and decline of civilization – but fortunately for the West, mainly progress. Objectively so. The objects speak for themselves.

Conclusion

Mayanism is not gone. Consider two of the most popular texts about the Maya in 2007: Jared Diamond's *Collapse* (a bestseller in the US two years after its release) and Mel Gibson's film *Apocalyto*. The former displays on its cover a stark black-and-white photograph of one of the great Maya cities, framed by one word, COLLAPSE. The latter, filmed entirely in Yukatek Maya, tells a story of imperial overreach (Gibson has said that he intends it as a metaphor for the US invasion of Iraq). Here again we find the morality tale about Maya civilization rotting out from the inside, destroying nature, and thus opening the door to European colonization. As if the point could be lost, Gibson begins his film with a quote with the pop philosopher Will Durant: "a great civilization is not conquered from without until it has destroyed itself from within."[112] By now it should be clear that these tropes are freighted by the ruins of Mayanist texts from which they gather their authority. Returning to Karatani, we may conclude that the "aestheticentrist way of appreciating and respecting the other" (well represented by Gibson's realist film style) that Mayanism teaches has never been a just or sincere mode of respect.

In the nineteenth century, Maya archaeology and Maya studies could not be separated from ethnography or development studies as contemporary disciplines. The three were united, in the case of Belize, with the support of the British Museum and the colonial state. As these disciplines evolved they separated over the distinction between the dead and living Maya: one became a science of bones and stones, the other a social science of culture and craniometry. In Mayanism, this distinction worked with a particular understanding of the relationship between the two, a disciplinary distinction that both divided and enabled. To place the "Maya world" required a discourse that could address the relationship between old and new Maya people and practices. When the challenge of finding, explaining, and appreciating the vestiges of ancient Maya life in and among the living Maya became defined as such, Mayanists both great (J. E. S. Thompson, Miguel Asturias) and amateur (Thomas Gann, Mel Gibson) have offered their works. Mayanism allows for the appreciation and explanation of the stories, clothing, and language of the living Maya. In so doing it places the Maya people within the territories of four nation-states, their practices within European reason.

Mayanism is therefore the European result of a double displacement of Europe. The first displacement is the discovery of Maya *humanity*. Las Casas is the figure we associate with this moment, when Europe confronts the humanity of the indigenous peoples of the Americas. This is a displacement, a decentering, insofar as it forces Europe to take account of itself in relation to this discovered other. And yet Las Casas' great contributions – his anthropology, proofs of the existence of reason, and faith in the potential salvation of the indigenous – amount to a fundamental *recentering* of Europe. It is a European judgment of reason and salvation that makes possible this assertion of human rights. Equally fundamentally, as we have seen *de Verapaz*, his contribution is to produce a hegemony of indigenous rights that clears space for a reasonable anthropology, a reasonable territorialization. The second displacement comes with the discovery of Maya *civilization*. The ruins and temples forced a new accounting of Maya as not simply people but fallen members of a great civilization. The knowledge of this moment is generated principally through Maya anthropology and archaeology, which have empirically uncovered the traces of ruins, histories, beliefs, times. Yet their contributions only *recenter* Europe: by measuring civilization in a world-time calibrated by Europe, defining the Maya as a fallen race, and territorializing the Maya within the space of nation-states. And from the ruins of Mayanism, capitalism qua development offers its settlement to the Maya.

Notes

1 Jones (1998).
2 Said (1994: 350). Pages 100–1 include language first published by Pion, Ltd., London, in Environment and Planning A, 2005, 37: 1033–43.
3 On this argument see especially Amin (1989); Mitchell (2000); Ismail (2005).
4 Said (2002 [1993]: 195). Gramsci's spatial emphasis is an effect of the challenge of explaining the production of hegemony in a non-economistic tenor. He sought to interpret the reasons that different social groups came to be separated or united as spatial subjects – and why they were often united by right-wing appeals to national belonging. In his unfinished essay on the "Southern question," Gramsci argues that the central problem for Italian communists is to articulate Italy's southern peasants with the northern proletariat. Gramsci saw that the Communist Party could not rely on the shared class interests of different social classes to generate a united front; the links between disparate social groups must be forged through concrete forms of political and intellectual leadership. The proletariat can only become the hegemonic class "in the measure in which it succeeds in creating a system of class alliances that will permit it to mobilize the majority of the working population against capitalism and the bourgeois state.... [In] the real class relations as they exist in Italy, [this] depends on the measure in which it is successful in obtaining the consensus of the large peasant masses" (1995: 19). The "southern question," in short, was how to transform the uneven terrain of knowledge and power on which the peasantry embraced fascism. Producing communist hegemony therefore implies reconfiguring the very spatial basis of political, cultural, and economic life in such a way that the proletariat and peasantry may realize their potential alliance.
5 Said's hermeneutics of the *Notebooks* are outlined in "History, literature, and geography" (2002 [1995]), where he argues that five themes are essential to interpreting the *Notebooks*: (1) the interconnection between political power and cultural life; (2) the opposition to determinism and economism; (3) the articulation of ideas and cultural practices with specific social classes; (4) the undermining of the theory-practice distinction; and (5) the intensely geographical tone. Of these, Said suggests that the last is the most profound, since it is Gramsci's "spatial sense of *discontinuity* that complicates and renders far less effective...the possibility of correspondence, congruence,

continuity, and reconciliation between different areas of experience"
(p. 458). This may warm a geographer's heart, but note that what is lacking
here is any reference to Gramsci's communism, and the very task of
producing communist hegemony that propelled Gramsci to conceptualize
spatiality and social practices in such an original and forceful way. Said's
reading of the *Notebooks* is certainly non-Marxist, and quite possibly anti-
Marxist; from them, Said gleans a conceptual repertoire for interpreting
spatiality, imperialism, and power, yet without addressing the question of
capitalism at their core. This lacuna has contributed to the distance
between postcolonial and Marxist approaches to imperialism and has yet
to be adequately addressed.

6   Ahmad (1992: 183). For an excellent rebuttal of the points raised in
    Ahmad's polemic, see Marrouchi (2000). I note in passing that this critique
    of Said has also been made by numerous conservative critics as well.

7   Said (1981: 4).

8   I borrow the expression dispensations of power from Anne McClintock,
    who claims "a *proliferation* of historically nuanced theories and strategies
    is called for, which may enable us to engage more effectively in the politics
    of affiliation, and the currently calamitous dispensations of power" (1992:
    97). Dispensation refers to the act of distribution, or dealing out; however,
    according to the OED, dispensation means also "the divine administration
    of conduct of the world," but also, notably, within law, "the relaxation
    or suspension of a law." Revising Agamben, we could speak therefore of
    "sovereign dispensation": only the sovereign (divine administration of
    conduct of the world) may suspend the law.

9   I take it that this is what Samir Amin (1989) had in mind when he wrote
    his critique of *Eurocentrism* – an explicit extension of Said's *Orientalism*,
    albeit one that places capitalism as a mode of production at the center of
    his analysis, and therefore an implicit critique of Said. In this light, my
    aim in this chapter could be seen as taking up Amin's proposal in his
    study of "defining [one of] the distinct arenas in which Eurocentrism is
    manifested" (viii).

10 Mayanism is also "American" because its emergence was facilitated by the
    rise of the USA as an imperial power in the nineteenth century: "When
    American [meaning here "those from the USA"] archaeologists, patrons of
    science, and the people at large shall take as much interest in this strange,
    purely American nation as they do in the ancient peoples of Egypt,
    Babylonia, and Greece, our knowledge and our admiration of them will
    be by so much increased" (Mason 1927: 380).

11 On the transnational indigenous rights movement, in Guatemala, Mexico,
    and Belize that is sometimes known as "pan-Mayanism," see Montejo
    (2005); Warren (1998). I am a partisan of the transnational movement
    for indigenous rights. Both Mayanism and pan-Mayanism must be dis-
    tinguished from the New Age "Mayanism" or "Reiki-Mayanism" that

consists of the celebration of Maya mathematical achievements (Alexander 2001). This "Mayanism" is not my concern here.

12 "Accomplices" is from Spivak. Empiricists call them "sources"; Marx once called them his "slaves."

13 Beyond the architectural grandeur of Maya history and the enigma of the collapse, there is also the relative geographical accessibility of Maya communities for American anthropologists.

14 Las Casas (1967), *Apologética historia*, pp. 128–9; cited in Keen (1967: 72–73), my italics.

15 Las Casas (1992 [1542]).

16 In 1898 an English translation of Las Casas' famous *Brief Account of the Destruction of the Indies* was published in New York under the title *An Historical and True Account of the Cruel Massacre and Slaughter of 20,000,000 People in the West Indies by the Spaniards* (Hanke 1953: 28). In 1898 the US invaded Cuba, Puerto Rico, Guam, and the Philippines, all previously Spanish colonies.

17 Hardt and Negri (2000: 115). It is this utopian element that keeps Hardt and Negri from celebrating "particularism and isolationism" as a response to imperialism, since the utopian qualities of people like Marx and Las Casas provide tools "to forge a project of counter-globalization [and] counter-Empire."

18 Widdifield (1990: 129).

19 Carozza (2003: 291); Griffin, in Las Casas (1992: xxvii).

20 Yáñez (2001: 131), my translation; Bolívar, cited in Carroza (2003: 296).

21 Moreiras (2000).

22 Las Casas (1992 [1537]: 63).

23 Ibid. My translation.

24 Moreiras (2000: 360). In Gramsci's terms, Las Casas was an organic intellectual: he provided intellectual leadership – an enlightened imperial reason – for a specific social class: narrowly speaking, this was the church itself (which was neither marginal nor subaltern in the Spanish Americas of the 1500s). More broadly Las Casas' intellectual leadership represented the ideology of the emerging capitalist elites of the Spanish American colonies, those like Bolívar who would later take up the cause of liberal-national development.

25 Ibid., p. 353, my italics.

26 Las Casas (1531), cited in Biermann (1971: 447–8).

27 Ibid., p. 453, my italics.

28 Hanke (1959: 29). Similarly, Biermann calls the Verapaz experience "the jewel in the chain" of Las Casas' work.

29 This was not Las Casas' first territorial experiment. Years earlier he attempted to materialize his theological arguments along the coast of Venezuela ("Tierra Firme"), but the experiement failed due to "Spanish encroachments on the coast" (Biermann 1971: 447). My discussion of the

Tezulutlán experience relies on Las Casas, Remesal, Bataillon, Biermann, Keen, and Hanke: see especially Remesal (2002); Bataillon (1971); Biermann (1971). The primary source for these texts is the account of Father Antonio de Remesal, a sympathetic biographer of Las Casas whose account has been criticized for its romantic portrayal of Las Casas and his careless treatment of historical fact. But for my purposes Remesal's writings on Las Casas are useful because, in memorializing Las Casas, Remesal dwells on the the practical connections between Las Casas' thought and the Verapaz experiment.

30 Remesal (2002: 89–90).

31 Ibid., p. 90.

32 Ibid., p. 94.

33 Ibid., p. 93.

34 Las Casas (1540): "Vuestra Majestad me mandaba que yo, con otros deligiosos de mi orden de Sancto Domingo, prosiguiesemos cierta pacificacion de muchas provincias que estan de guerra, trayendolas al servicio y subjecion de Vuestra Majestad."

35 Bierman (1971: 478–9), my italics.

36 *Reducción* is typically translated as "resettlement," but as we have seen, the Spanish did not necessarily see the Maya *as* settled peoples. "Forced relocation" would seem more appropriate. In Remesal's account of the *reducción* of the Maya communities of the Verapaz, the Chief is said to have agreed with Las Casas' principles, and so he "surveyed the territory" to "put the plan [*reducción*] into practice.... But then he encountered opposition, and soon some took up arms, disliking the idea of abandoning their isolated hovels in the hills, canyons and valleys, where they were born. This greatly afflicted the two padres who lost some of the goodwill they had enjoyed.... But the Lord supported them and slowly they managed to create a community of a hundred houses, under the name of Rabinal" (2002: 100).

37 Remesal (2002: 99–100).

38 Ibid., p. 105.

39 Agamben (1998: 15).

40 Ibid. Agamben writes: "The particular 'force' of law consists in this capacity of law to maintain itself in relation to an exteriority. We shall give the name *relation of exception* to the extreme form of relation by which something is included solely through its exclusion" (p. 18).

41 Gramsci (1971).

42 On the three conquests of the Maya, see Lovell (1988).

43 Moreiras (2000: 353), my italics.

44 Las Casas elaborates his arguments especially in his *Brevísima relacion de la destrucion de las Indias* (Seville, 1552) and the *Historia de las Indias* (Seville, 1875 [1561]).

45 The first conquest of the Maya was led by the Spanish conquistadores in the sixteenth century. The second conquest of the Maya of Guatemala was

in the latter half of the nineteenth (e.g., most of the ancestral lands of the Q'eqchi' were stolen during the coffee boom of the 1870s). A third round of conquest was launched in the late 1970s, when conservative forces pushed through new demands for Maya lands in the Alta Verapaz. When hundreds of Q'eqchi' Maya peasants met in Panzós on May 29, 1978 to protest plans of state leaders to expropriate their lands, they were met by a hail of bullets; the death of more than a hundred unarmed Maya peasants, the massacre at Panzós, Alta Verapaz, marked the onset of the Civil War that ended in the death of roughly 300,000 people. Except that the war has not really ended: the 1996 Peace Accords have done little to change the massively unequal distribution of power and land in Guatemala, where 65 percent of arable lands are held by only 2.6 percent of the population (see Black 1998: 13).

46   Las Casas (1992 [1537]: 65).
47   Cf. Keen (1977).
48   I am indebted to Qadri Ismail (2007) for this argument.
49   Las Casas (1967: Vol. 1, p. 648).
50   Landa's account of the colonization of Yucatán was not written until 1566, and not widely read until after the 1860s. Moreover, its provenance is uncertain: see Restall (2003).
51   Hardt and Negri (2000: 116), italics mine.
52   Gallenkamp (1976: 191).
53   Morley (1946: 141).
54   Coe (1977: 46).
55   One of the important aspects of their relationship, in terms of their popular success and textual authority, is that they were from the US (Stephens) and England (Catherwood); see von Hagen (1973).
56   Stephens, cited in Gallenkamp (1976: 46).
57   Stephens (1963: 165).
58   Hay sponsored a major Egyptian expedition that included many famous orientalists, including J. G. Wilkinson, G. Hoskins, F. Arundel, and Edward Lane. In Egypt he drew and mapped Cheops, Chephren, and Mycerinus. "The result of the expedition was forty nine large-folio volumes furnished with hundreds of pictures that Hay gave to the British Museum in 1879" (von Hagen 1973: 24). When Catherwood worked in Egypt, he entered the Omar Mosque dressed as an Egyptian official: "Showing uncommon determination and courage, he became the first 'infidel' to set foot in the fabulous and splendid Omar Mosque. He spent some days beforehand sketching the exterior of the mosque, then, on 13 November 1833, dressed as an Egyptian official, he could not resist the temptation to draw the interior" (ibid., p. 28).
59   This text was published under the pseudonym "An American." *Incidents of Travel in Egypt, Arabia Petrea and the Holy Land* was a commercial success in the US and Europe.

60 Catherwood also published a collection of his drawings in *Views of ancient monuments in Central America, Chiapas, and Yucatan* in 1844 (London: Archivio White Star).

61 The decline of civilization is thus symbolized by the movement from a symbolic to an organic relationship with the snake. The snake, of course, is hardly an innocent sign, but the very trope of the fall into worldliness, knowledge, and sin (Genesis 3: 1–24).

62 Wilk (2002: 4). For a recent and provocative example of this Malthusian fable, see Diamond (2003: 51). Diamond reiterates Cook's argument that the ancient Maya "grew corn by means of a modified version of swidden slash-and-burn agriculture, in which the forest is cleared, crops are grown in the resulting clearing for a few years until the soil is exhausted, and then the field is abandoned" (46). The editors of *Harper's* smartly juxtapose Diamond's essay with five illustrations by Stan Fellows that mock the essay's essentialist, Malthusian tone, by melding Maya glyphs with symbols of American militarism. One glyph shows a Maya warrior running to battle underneath a cow sign; one has a stealth bomber behind a Maya priest; another, a Maya farmer planting maize, a biohazard sign on his *kuxtal*. The map of the "Maya area" includes Crawford, Texas.

63 In MP 3267 of 1894, Frederic Keyt, the District Commissioner of Toledo, writes to the Colonial Secretary concerning the discovery of ruins on the Rio Grande. The ruins are said to consist of "temples and many buildings, covered with vegetation and high bush." No official report was prepared. In MP 3442 of the same year, Keyt wrote to the Colonial Secretary reporting that he had "established existence of Maya Ruins." Keyt posted a warning in four of the stores in Punta Gorda against the damaging of ruins.

64 Hammond's account of the history of archeological work at Lubaantún notes that Gann's "enthusiasm was not matched by his expertise in either excavation or observation" (Hammond 1972: 7), which is a kind way of saying that Gann's techniques involved dynamite and guesswork.

65 Mayanist research has been tightly linked with British and US politics towards the region from the outset. Stephens was treated as a special envoy to the President of the US.

66 Gann (1925).

67 The popular travel-writing mode is betrayed by the short summaries at the introduction of each chapter. Chapter 11 of *Mystery cities* promises: "Two Maya dialects spoken around Lubaantun – Indian's mode of life little altered in last three centuries – Old village sites frequently found in the bush – Difficulty in junking great trees, too large to burn . . ." (Gann 1925: 163).

68 Gann Thompson (1937).

69 The *Illustrated London News* reported on July 26, 1924 that Gann "discovered" Lubaantún, not noting that he had already written about the site twenty years earlier. Hammond plausibly speculates that Lubaantún's

"rediscovery" may have been the "result of the 1924 British Empire Exhibition" (1972: 12). This would not be the first time that archaeology was pressed into the service of colonialism.

70 Gann (1925: 134).

71 Ibid., p. 135.

72 Ibid., p. 136.

73 Ibid., p. 182.

74 Ibid., p. 178. Can we imagine a better metaphor for the racist teleology of social evolutionary thought than "the ladder of degeneration"? (Gann introduces many of his most racist and unsubstantiated claims with the expression, "in fact. . . . " Empiricism here is shown to be collaborating with colonial racism.)

75 Ibid., p. 156.

76 Ibid., pp. 156–7.

77 Lundell (1933: 70).

78 This section is derived from research conducted in collaboration with Dr. Josh Lund. An earlier, co-authored version of this argument is forthcoming in *Interventions: A Journal of Postcolonial Theory*. I thank Josh Lund, *Interventions*, and Taylor & Francis for permission to use some of our material here.

79 See Martin (1993).

80 Asturias (1993: 11).

81 Ibid., pp. 9, 220.

82 "Asturias's engagement with the actual pages of the Maya masterpiece *Popol vuh*, as a student of Georges Raynaud, demonstrably did more to change his view of the Maya than did any other experience of the time" (Brotherston 1996: 431).

83 The Maya were first colonized by the Spanish, but the most celebrated Mayanists before Asturias's time were British, such as Thomas Gage, Frederick Catherwood, Alfred Maudslay, and Thomas Gann. Many of the objects that Asturias would have studied at the British Museum would have been carried to London by Maudslay and Gann.

84 Harss and Dohmann (1967: 76).

85 In his biographical chronology of Asturias, Martin writes that in 1903 (age 4), in Salamá, Baja Verapaz, Asturias "entra en contacto por primera vez con los indígenas de su país" (1993: 461). Later, beginning in 1908 (age 9), on the patio of his father's trading warehouse, "el joven Asturias pasa muchas horas conversando con arrieros y campesinos" (ibid.). These influences, along with his indigenous nanny, are usually taken by Asturias's readers as a prominent and direct link to the living voice of the Maya. See Henighan (1999: 198–9) for a reading of the limitations of these sources.

86 Asturias (1977: 65).

87 Ibid., p. 105.

88   Ibid., p. 61. The logic works upon three premises. First, "territory" is "our piece of the globe." Second, "its population" is the "people that inhabit" that territory. Third, Asturias says that the central concern is the relation between the two, i.e., his study takes as its object the articulation of the *nation* with its *territory*. This crucial articulation will remain with Asturias throughout his writing life. The problem is understanding and reforging this articulation, since the colonial method has failed.

89   Asturias (1993: 72).

90   Ibid., p. 237, my italics. ("Vas a comparar vos lo que eran antes estas tierras cultivadas por ellos racionalmente...El méiz debe sembrarse, como lo sembraban y siguen sembrando los indios, para el cuscún de la familia y por negocio...Y todo este cultivo tenían los indios...en pequeño, si vos querés, pero lo tenían, no eran codiciosos como nosotros, sólo que a nosotros, Hilario, *la codicia se nos volvió necesidá*": 219–20, my italics).

91   Asturias, 1967, "Nobel Banquet speech," cited at www.nobelprize.org/literature/laureates/1967/. My italics.

92   Asturias (1993: 191–3).

93   Ibid., p. 304.

94   On the primitive accumulation of Maya land in Guatemala, see Galeano (1967); Melville and Melville (1971); Handy (1984); McCreery (1994).

95   Asturias (1993: 11).

96   For Rivera's treatment of wheat, see the *Capilla Riveriana* at the Universidad Autónoma Chapingo (Mexico's national agricultural university), where maize and wheat stand in counterpoint as symbols of the American and European agricultural traditions. At least in Guatemala, wheat makes for a curious symbol of imperialism. True, it has partly displaced maize as a staple, but unlike bananas and coffee (the crop that spurred the liberal reform), wheat is not a plantation crop produced for export.

97   Harss and Dohmann (1967: 75), my emphasis.

98   Asturias (1973: 6). The thought that Maya maize farming has not changed is incorrect. To be fair, in Asturias's day it was widely surmised that ancient Maya maize was "sewn and reaped [exactly as it is] today." In this sense Asturias does not break with the predominant Mayanist discourse. He extended the argument to stress the *value* of the *milpa*. It lay at the center of "this great [Maya] culture, alive in its roots, [which] must be included in the current dialogue of cultures – with its message of beauty, its human dimension, and its perennial rebirth" (p. 9).

99   On the British Museum's support for these and other Mayanists, see Carmichael (1973).

100 Latour (1987: 215–57).

101 The British Museum opened in 1759 and had acquired many of its best-known objects by the 1820s: "Governed by a body of Trustees responsible to Parliament, its collections belonged to the nation....The first

antiquities of note, Sir William Hamilton's collection of Greek vases and other classical objects, were purchased in 1772. These were followed by such notable acquisitions as the Rosetta Stone and other antiquities from Egypt (1802)...the sculptures of the Parthenon, known as the Elgin Marbles (1816)....In 1823 the gift to the nation by George IV of his father's library (the King's Library) provided the catalyst for the construction of today's quadrangular building designed by Sir Robert Smirke. The first phase was largely completed in 1852, to be followed by the round Reading Room...erected in the central courtyard in 1854–7." Accessed from the British Museum website, October 2006, at: www.thebritishmu seum.ac.uk/visit/history.html.

102   House of Commons. 1999. "Memorandum submitted by The British Museum." House of Commons, Select Committee on Culture, Media and Sport Minutes of Evidence. Accessed November, 2006, from: www.publications.parliament.uk/pa/cm199900/cmselect/cmcumeds/371/0060802.htm.

103   On the marbles, see Hitchens (1998).

104   Accessed from the British Museum website, October 2006, at: www.thebritishmuseum.ac.uk/compass/ixbin/hixclient.exe?_IXDB_=compass&_IXFIRST_=1&_IXMAXHITS_=1&_IXSPFX_=graphical/gt/int/top_&$+with+all_unique_id_index+is+$=ENC13300&_IXtour=ENC13300&submit-button=summary.

105   Ibid.

106   Slessor, C., "Mexican marvels: New Mexican Gallery in the British Museum," *Architectural Review* (January, 1995). Downloaded October 2006 from: www.findarticles.com/p/articles/mi_m3575/is_n1175_v197/ai_16565505.

107   The text by the head in the museum says that it was "given by A. W. Franks, Ethno +4076; bequeathed by Thomas Gann, Ethno 1938." I attempted to reconfirm the provenance of this jade head at the British Museum and was told to write a letter to the Director of the Department of the Americas. I did so but received no reply.

108   This is a play on Althusser's famous thesis on interpellation: "ideology 'acts' or 'functions' in such a way that it 'recruits' subjects among the individuals (it recruits them all), or 'transforms' the individuals into subjects (it transforms them all), by that very precise operation which I have called *interpellation* or hailing" (Althusser 1971: 174).

109   Heidegger (1977a). Heidegger did not, to my knowledge, acknowledge or examine the affiliation of enframing with state power in territorialization, or the relation between enframing and imperial power. But see Mitchell (1998).

110   This can be seen not only through European museums generally, but also the original room of the British Museum: a hall of curiosities, with bits of every ancient culture represented by a jumble of artifacts, sorted

not geographically but thematically (religion; science; etc.). Quite appropriately, today this room is organized by the theme, "European Enlightenment."

111 House of Commons. 1999. "Memorandum submitted by The British Museum." House of Commons, Select Committee on Culture, Media and Sport Minutes of Evidence. Downloaded November, 2006, from: www.publications.parliament.uk/pa/cm199900/cmselect/cmcumeds/371/ 0060802.htm.

112 Downloaded from www.apocalypto.movies.go.com/. For contemporary Mayanism see also Gugliotta, 2007.

Part II

Aporias of Development

4

From Colonial to Development Knowledge: Charles Wright and the Battles over the Columbia River Forest

I do not intend to deconstruct the colonial discourse to reveal its
ideological misconceptions or repressions, to exult in its self-reflexivity,
or to indulge its libratory "excess." In order to understand the productivity of colonial power it is crucial to construct its regimes of truth, not to
subject its representations to a normalizing judgment. Only then does it
become possible to understand the *productive* ambivalence of the object
of colonial discourse – that "otherness" which is at once an object of desire
and derision, an articulation of difference contained within the fantasy of
origin and identity.

Homi Bhabha (1994: 67)

Is it possible for an agent of colonialism to oppose empire? If we read
Orwell's dispatches from Burma, it seems that there is an anti-imperialist
spirit beating in the heart of every colonial officer – at least, perhaps,
among the subaltern ranks. And yet criticism by agents of colonialism is
always marked by certain limitations. As a founding condition of their
criticism of empire, most colonial agents clear the space of their critique
by defending the innermost quality of imperial practice: rule from outside. This problem often manifests (as Said has shown in his readings of
Conrad) in the inability to recognize, in the midst of criticism of empire,
the possibility of an autonomous, indigenous critique that is not always

already translatable in terms reserved for the "other."[1] Thus even when criticizing empire, the critic continues to speak for and silence the colonized.

In this chapter I explore the dynamics of this problem through a reading of the enigmatic works of A. C. S. Wright. A soil biologist in the service of empire, Charles Wright wrote the land use plan for colonial Belize – for which he was honored with the Order of the British Empire – and went on to purchase land and establish a farm in southern Belize. But Wright was no mere land-use planner or tropical farmer. He was, I argue, the most important colonial agent in southern Belize in the twentieth century. His best-known contribution, *Land in British Honduras* (hereafter, *Land*), set the stage for later development practices in the region.[2] As Wright wrote forty years after it was published: "During the years leading up to independence, when British Honduras became Belize and Colonial Administration was gradually phased out, the recommendation[s] of the Land Use Survey Team were broadly followed."[3] And much as John Taylor's texts marked a shift in colonial discourse (see chapter 1), Wright's work in the 1950s serves as a hinge around which the discourse about the Maya and development turns; through *Land*, the discourse of Maya agricultural development is crystallized as a form of modern, *disciplinary* knowledge. After 1959, all writers on Belize cite *Land*, and very few go any farther back into the archive. *Land* stands not only at the beginning of development, but also as a bulwark against an older form of colonial discourse.

This shift is related to a broader pattern. Arturo Escobar and others have argued that the discourse of development was essentially created anew on a global scale in the postwar period.[4] While Escobar is correct in noting that development discourses underwent a profound shift in this period, I find that his thesis is too general, too geographically undifferentiated, to allow us to understand the complexity and dynamics of the colonialism-development relation. Perhaps it is because Escobar treats development as something "constructed" *de novo* after World War II that he calls for us to simply do away with it. I argue that if we aim to decolonize development, we must treat "development" as something other than an all-pervasive and undifferentiated ideology. We must carefully examine the ways that particular colonial tropes were variously reworked through development discourses – and investigate the effects of these discourses on the production of spaces, environments, and livelihoods.

This requires a consideration of the work of Wright. What makes Wright an especially complex figure for our purposes is that in 1967 he "retired," in his words, to southern Belize. He bought land and became a new kind of colonist during the period of decolonization – one devoted with a singular focus and commitment to understand, document, and transform rural Maya livelihoods. In a word, he was an *applied Mayanist*. His work was polyvalent and passionate to the end. Wright lived in Punta Gorda for the last three years of his life (1995–8), surviving as a consultant and enjoying the most productive period of his life as a writer.[5] During this period Wright enjoyed wide popularity, even fame, in Belize. His room in the Belize Center for Environmental Studies was a veritable salon of researchers and friends who came for stories of colonial times, lessons on the Belize environment, and a taste of rum. I was one of many students who came to learn from him.

Despite his efforts to distance himself from empire, to downplay his political involvement, his stories and prodigious environmental knowledge were rooted in his years of research experience as a colonial soil scientist. And as we will see, Wright remained a spokesman for Belize's natural resources in ways that retained an imperial tone. Nonetheless, in his later years Wright became widely known as an advocate for the Maya and their forests. The best-known colonial critic of the Maya farm system in the 1950s – the natural scientist most responsible for soliciting development practices – became one of the best-known advocates of both forest conservation and Maya land rights. This irony raises the question of what Bhabha refers to as the "productive ambivalence of the object of colonial discourse": how representation, power, and "speaking-for" become entangled in the itinerary of colonialism and development.

Edward Said argues that a biographical approach to literary criticism cannot but produce a unified subject and a narrative that rests upon a linear, unfolding temporality: "Identity, which is non-contradiction, or rather contradiction resolved, is the heart of [biography], and temporality its sustaining element, the essence of its constitutive structure."[6] I recognize that I run this risk in this chapter, and guard against it by emphasizing Wright's texts. My aim is emphatically not to define the essential Charles Wright, nor write a history with him as subject. For our purposes, the pertinent facts of Wright's life can be briefly summarized: he was raised in a comfortable, intellectual atmosphere in Cambridge, England, and encouraged to pursue the sciences.[7] After graduating from Leeds University with honors in biology in 1936, he went to New Zealand on a Commonwealth scholarship and conducted soil and ecological surveys for four years. After service in World War II, he became Chief Soil Biologist for the New Zealand Soil Bureau, from whence he was recruited to head the Land Use Survey Team in 1952. He spent most of his remaining years in southern Belize, and by his death in 1998, wrote scores of texts on Maya land use and agriculture.[8] These texts, and not Wright *per se*, are the object of this chapter.

From Colonial to Development Knowledge: *Land in British Honduras*

We must begin with *Land in British Honduras*, Wright's first and most important text about Belize. We should not be thrown off by the way the text represents itself – as the report of an objective soil survey. No doubt, it is a report of a soil survey, but also more; the soil comprises the text's O horizon. Above all, *Land* constitutes an attempt to forge a disciplinary approach to the question of development solicited by the colonial examination of the Maya. As we have already seen, the space for this project was cleared by the colonial discourse on the Maya farm system. *Land* therefore stands as a turning point in the history of discourse about development and the Maya in Belize. It is unimaginable without the preceding texts on colonial management and the Maya farm system that called forth its problematic. Yet it also breaks with these in one crucial respect. Whereas the earlier texts formed a constellation of new concepts, they had not arranged these in such a way that prescribed correlate forms of development trusteeship; in Cowen and Shenton's terms, they had not been crystallized as development doctrine. Without reiterating the arguments of Part I, we may briefly locate *Land in British Honduras* in the archaeology of Mayanism in these terms: after the consolidation of the Maya farm system as an object of discourse, *Land* reconfigures the Maya in relation to capitalism qua development.

The conditions of production of *Land* reflect its colonial mission. In 1933, one of the Rangers of the new Forest Department reported on the "Indian occupation of the Toledo District," producing the first maps of the Indian reservations and the "origins" of the Maya in Guatemala.[9] Reports of suffering in the Maya communities during the 1930s – given voice mainly by church fathers (see chapter 1) – reached London and the Secretary of State for the Colonies called for an investigation by an "Interdepartmental Committee on Maya Welfare" (ICMW). The 1941 ICMW report is notably the first text to define colonial development policy for the Mayas as "replac[ing] the milpa system"[10] and for proposing that Maya welfare would be improved through the scientific study of the colony's resources. The Land Use Survey Team was formed to take up these challenges in August 1952, when Wright arrived in Belize as a veteran of World War II and the New Zealand Soil Survey.[11] The research for the text was conducted between September 1952 and February 1954, with funding from a Colonial Development and Welfare grant.[12] After the initial results were presented in Belize in 1954, the four British team members retired to the Imperial College of Tropical Agriculture in Trinidad to write the book.

   Land is comprised of three sections. Part I reviews the geographical, climatological, ecological, and cultural "patterns that will influence land use in British Honduras,"[13] and Part II provides the results of the soil survey. Part III, "Developing the soil resources of British Honduras," is the longest and most important for our purposes – but before we turn to it, a word on the text's structure. *Land* describes geographical patterns (climate, ecology, culture, etc.), moves into the detailed description of soils, and then says how development should proceed. The territorialization of Belize and the expectation that it is to be developed as a set of national resources are effects of this structure, which implies that the prescriptions for development merely extend from the analytic description of the colony's territory. From the unproblematized ground of geography (I), to the interrogation of soils (II), to the conclusions about development (III), there is never an attempt to reflect upon the relation between the text and its effects, nor to justify this approach. It is posited as self-evident. Nor do the authors support the narrative through appeal to references, as one might expect from a scientific text. *Land* refers only sparingly to other texts. The literature review comprises eight sentences out of 327 pages, and there are few references elsewhere.[14] There is no bibliography. Whenever the argument is strained, the narrative returns to the soils – as though *Land* sprang forth from the land itself.[15]

Developing the Maya Farm System

Land works with a particular object of development: the *milpa* farm system. This is, *Land* says, "the basic unit for production" in Belize.[16] How it came into existence as an object *for* development is not examined. The question is simply how to develop the *milpa* farm system given existing constraints. *Land* initially poses this as a challenge of understanding: "The development of British Honduras really starts with *understanding the 'milpa'* and studying the way of life of the 'milpero.'"[17] Development therefore calls for applying knowledge in such a way that *settles* the *milpa*: "A solution to the wandering 'milpa' has to be worked out and permanent farming introduced."[18] Although Wright et al. find variations on the *milpa* throughout Belize, they argue that development should focus on the rural south because this region has the lowest level of development, high-quality soils, and Maya farmers (the latter are said to be "skillful," "energetic," and "adaptable"). Although the Maya are treated as a race with distinct problems (as we will see), they are also praised as "unquestionably the most industrious group of farmers in the whole country and a group with very long farming traditions."[19] Hoping to capture the power of these traditions, *Land* places "progressive" Maya farmers at the center of an agrarian reform that would incorporate the Maya into the national economy.[20] In other words, the Maya and their farm system are treated as a great *resource* for development – since the Maya are the only farmers in Belize with longstanding agricultural traditions. The Maya farmer "represent[s] the most valuable agriculture asset in British Honduras, worth a dozen overseas investors, development corporations and the like."[21]

But the Maya farmer is not an asset for himself. He is an asset – a resource with potential, not unlike capital – for the *colony*. Therefore it is in the interest of the colony for *Land* to propose a strategy for the reform of this object. But the Maya farmer is an object marked by profound ambivalence. On one hand, he is appreciated as an "industrious," "intelligent," and "adaptable" farmer who achieves a relatively "high level of production."[22] On the other hand, he is "lacking in modern farming techniques," a "poor mathematician," and he moves too often; his farm system depletes soils and forest land;[23] and he is isolated from the real life of the colony. His character is questionable, too, as he tends to "throw away the labour of weeks" whenever he sells his corn.[24] These intense ambivalences frame *Land* in three important ways.

First: they coordinate the resource development plan for the colony through a discourse of *race*. At every point when *Land* examines the relationship between land and the Maya, we find an ethnographic tone. For instance, about "the Kekchi" we read:

> He brings to the common pool a habit of independence and self-dependence and the ability to make a careful analysis of cause and effect whenever he encounters a strange phenomena [sic]. The Kekchi have, in addition, a fine sense of humour and the gift of pantomime. They are progressive farmers, quick to see and adopt useful ideas. The Kekchi is also something of a wanderer and will wander to Guatemala in May, June, and July looking up relatives in that country.[25]

These statements may seem unbearably arbitrary, yet they are linked to a broader Mayanist discipline and they produce important effects. "The Kekchi" here is treated as a member of the Maya *race* which is marked by essential qualities (said to vary between the different "sub-races"). It is only as a *Maya* farmer – with his ancient connections to these soils – that the farmer constitutes a resource for the nation-state. Such essentialist distinctions inevitably exclude certain subject-positions from being recognized and fix the place of each racial group in the colony's development by arbitrary and orientalist designations.[26] It is only the Maya man (always "he") that *Land* treats as a resource.[27] He is radically unsettled: a recent in-migrant to the colony, he is said to have longstanding ties to the land – not historically, but by virtue of his affiliation with the ancient Maya; however, his "link" to the ancient Maya is found to be "tenuous."[28] Unsettled in relation to his own place in history, he is "something of a wanderer" by his nature – and his *milpa* wanders, too.[29]

Second: the ambivalences around the Maya farmer frame development as a project of reconstituting colonial trusteeship. Insofar as the Maya farm system is a resource for development, it is an object with great *potential*. Earlier we saw that *Land* treats the Maya farmer as an asset. To develop "the latent possibilities of the Maya Indians" requires the integration of the Maya into the life of the nation: "Every effort should be made to bring the younger generation of the Mopan and Kekchi Indians into closer contact with every-day life in British Honduras."[30] This will require, *Land* says, the "[e]mancipation of the Maya Indians and the integration of these fine people with the other racial groups of the country."[31] For its value to grow and circulate, the Maya farmer as a racialized resource must be liberated. Who is the agent of emancipation? Development as facilitated by the colonial state. Thus, the identification of the Maya farmer as an unintegrated, incomplete resource brings about his *resubalternization* as a subject of development. If the Maya farm system was created to be an object of development, the Maya farmer was produced as a subject under colonial trusteeship for capitalism qua development.[32]

Third: the ambivalence around the Maya farmer articulates colonial anxieties about population with the project of producing territory. The discourse of "developing the Maya" binds the nation and the state through territory as a space of resources. Consider the discussion of overpopulation in *Land*. Speaking of the Maya farmer's use of land, Wright et al. argue that each village "grows every year by about 10 extra farmers, each of whom will want to use at least 10 acres of good bush land a year for the next five years, so that the annual need for new milpa land is certainly not less than 500 acres a year."[33] (I note in passing that this calculation presupposes that each farmer will need ten additional acres each year for five consecutive years; Wright et al. offer no evidence for this claim. Its dubious veracity is less significant than its tone and position within the narrative.)[34] The authors assert that the Mayas of southern Belize are growing too fast and demand: "Where is [more land] going to be found?" They posit that expanding the size of the Indian reservations will not solve the problem of population growth, since doing so would perpetuate the separation of the Maya farmer from the life of the colony. Allowing the Maya to use more land "would be shirking the real problem," they claim, albeit without specifying what comprises "the real problem." The only grounds for solving the population problem are settlement, development, and national integration.

How could the Land Use team pose themselves as arbiters of the appropriate distribution and density of racial groups in the colony? From what position do Wright et al. determine what comprises a race, and how do they know that "separation" is a problem? Herein lies the work of colonial epistemology: as *Land* is a scientific report of a survey of a colony – a member of the Commonwealth family – the authors are positioned to confidently see the different family members within this space. In the preface, Wright et al. explain how such familial ties inspired and held their team together:

> When the rain fell and the muddy trail seemed endless, we were sustained by the thought that we were working on behalf of a still younger member of the [Commonwealth] family; a country that needs technical and financial help as badly as any of the areas for which Britain has made herself responsible for progressive development.[35]

The clause "made herself responsible" provides a clue to the relationship between these Commonwealth family dynamics, the desire for integration, and development. The colonial desire to bring the family together and to provide for its welfare is felt strongest by the parental figure. *Land* produces the coherence of this territory for a nation-state – within the Commonwealth. One of the most important effects of *Land* is the thematization of soils, cultures, and trees as *national* resources. They are national and natural objects endowed with colonial worldliness.

The implications of such a geography are reflected in *Land*'s treatment of the question of land scarcity. Despite the fact that population density in the 1950s in Toledo was under four people per square mile, land scarcity is presumed to be a problem in the Maya communities. The solutions to the problem of land scarcity – like the question about where "this land" could be found (who is responsible to find this land?) – are posed within a colonial purview that seeks to fix the *milpa* within the space of the nation-state. Thus Wright et al. conclude that the solution to the problem of the overuse of land by the Maya farmer is to fix him for once and for all: "The wandering 'milpa' must be pinned down."[36] Reiterating the language of the Interdepartmental Committee on Maya Welfare, they call for "a more permanent type of farming to supplement and later replace shifting cultivation [with] the ultimate objective for the Maya Indians [of] complete integration in the life of the country." Whereas in the late nineteenth century the colonial desire to fix the "Indians" led to the creation of the reservations, for Wright et al. fixing the wandering *milpa* requires the radical *integration* of the Maya into the life of the colony. This means the Indian reservations must "be abolished... by settling every family on a plot of land."[37] (Without any nostalgia for the colonial history of the institution, we must acknowledge that the Indian reservations have had the effect of keeping 77,000 acres of land in southern Belize off the land market. These lands have been collectively governed by the communities through the alcalde system: see chapter 5.) So the anxiety about overpopulation, coupled with the logocentric desire for a settled, pinned-down Maya subject, leads to a call for the abolishment of the reservations, the privatization of the commons, and "integration" with the country.[38] Mayanism and soil survey thus produce this truth: the *colonial state* should free the Maya farmer – through *development*.

These three effects of the ambivalences around the Maya farmer in *Land* can be read, therefore, as specific effects of a development discourse that has some fairly broad parallels with the forms of development discourse emerging throughout the British colonial world. Some of these elements are implicated in effecting trusteeship and territorialization. For instance, the emphasis on the "emancipation" of the Maya farmer into the life of the nation is widely recognizable in national-populist development discourses that equate modernity with the abstract liberation of the rural poor. Thus, Wright et al. argue that development will not begin until farmers "find out that FARMING PAYS."[39] Yet they do not investigate the causes of low food prices or the broader political-economic conditions that have led Belize to be a food-importing economy. In short, they do not consider the relationship between colonialism and the agricultural economy. They simply posit that "what is needed most" is "energy, enterprise, inspiration, and sincerity."[40] The text not only claims that inspiration and sincerity are the keys to development – it seeks to *provide* it, in its enthusiasm, capitalization, exclamations, and tone. *Land*'s textual performativity thus articulates the populism of a farmer-first strategy with the elitism of a trustee's view.

Land as Development Doctrine

> From a land use point of view the Maya Indian – the indigenous user of land in British Honduras – is going to be worth all the trouble that we may take to advance his skill as a farmer.
>
> Wright et al. (1959: 285)

In chapter 2 we saw that the discourse of the Maya farm system was consolidated in the 1920s and 1930s. Now we can add that in the subsequent two decades, this discourse solicited capitalism qua development as a theme appropriate to new forms of trusteeship.[41] That is to say: "development" was not simply invented as a kind of free-floating ideology, or "constructed" all at once on a global scale in the postwar period. Rather, at least in this period and in this region, development discourses were reorganized and concentrated – through ways of framing the prospects for the Maya.

The effects of such framing in *Land* are especially powerful because of its disciplinary qualities: the soil maps, measurements, ethnographic descriptions, and prescriptions for change comprise a disciplinary ensemble. I stress that this ensemble always speaks in terms of development *for* the Maya. Populism is amenable to trusteeship. Wright et al.'s position here is clearly elitist – as reflected in their insistence in speaking for – and it is woven through a populism that treats the cultural knowledge of the Maya farmer as the resource at the center of development:

> Someone connected with the [development] plan must get to know the people intimately. Without this knowledge much time and money will be wasted later when Government seeks to implement the plans. [...H]e needs to know the right time and the right way to introduce a new project to the farmers. He needs...to become skillful in applied psychology.[42]

We have reached a point where a new figure is materializing before us. Who is this "he" who will guide development – with a psychologist's sensitivity to the character of the Maya farmer? Wright et al. refer to "a coordinating officer" in passing. We should read this not as the name of an occupation, but rather as a subject-position – one that is materialized by the discourse of development at this particular moment. And we should track this subject-position in the institutional materiality of the state's development practices.

One of the first projects materialized as a consequence of the Land Use Survey was a land scheme in the Maya village of Crique Sarco. Consistent with their discourse about the need to pin down the Maya farmer, the project aimed to create landed, propertied, family farmers in one rural Maya community. Wright explains the story behind the scheme in this way:

> As far back as 1952, there was already a strong feeling in the Colony that Mayan farming would eventually have to be changed from "shifting agriculture to permanent agriculture." The Team did a little extra research on this topic on a parcel of Crown Land right across the Temash River from the small Indian Reservation allocated for the villagers of Crique Sarco. The latter complained that their legal reservation of about 1,500 acres would soon be too small for the twenty families of Crique Sarco village. From the Colonial authorities in Belize City the Team obtained permission for an extension amounting to 2,000 acres of adjacent land just across the river.

I interrupt the story to briefly remind us of Wright's question about the problem of land stress: "Where is [more land] going to be found?" Here is a case where Maya farmers wanted to expand their *milpas* outside of "their legal reservation." The farmers were conceded 2,000 acres by the colonial state for them to use because of Wright's application for permission. Wright added the requirement that the land would be privately owned and parceled. Colonialism has so deeply normalized the relation between Wright and the Maya that it may not seem questionable or arbitrary that he would do so, yet we should ask: upon what ethical or historical basis does Wright play this role? How did Wright come to have the right to abrogate this land and determine the conditions of its distribution? These questions are not raised in *Land*. As for Wright's question (where will the Maya get more land?), he answers it himself, saying, in effect: when the Maya need land, let them come to me. The colonial state will give the Maya the land. This relationship of provision and responsibility is what constitutes trusteeship. We continue:

> While working in the area the Team surveyed, cut, and marked out twenty farms of 100 acres each and distributed these amongst the twenty families by ballot. The Team also surveyed the soils on each family's section and designated suitable methods of land use for each kind of soil. For each

farm there were soils alongside the river suited for coconuts or citrus trees; followed by a strip of flat land with soils more suited for rice growing; thence, moving away from the river, was a strip of soils suited for milpa crops or pasture for livestock, and for a house site; next came a strip of rolling to hilly soils well suited to corn and other annual crops; finally, at the back of each section, there were soils well-adapted to cacao, coffee, nutmeg, and some other tree crops. The villagers enthusiastically helped with the survey work without cost, and each family started to accumulate house building materials.[43]

Although the labor for the project was volunteered by the residents of Crique Sarco, it was felt that the project needed oversight from a British development officer. To this end, Wright asked the colonial government to provide his team with a "Liaison Officer" who would serve as a conduit between the Maya communities and the colonial government. They were joined in 1953 by the first and only Liaison Officer, Don Owen-Lewis, who moved from England to Crique Sarco to supervise "the development of the new farms." Notwithstanding the presence of the Liaison Officer, the experiment ended "a complete failure," as Wright explains:

> For the first three years all went well, but the news had by then spread far and wide (especially amongst Mayans living in Guatemala) that whole families could claim kinship (however remote) with the lucky owner of a whole block of 100 acres at Crique Sarco, who then became inundated with relatives expecting to share his good fortune, well-knowing that such a request would be very difficult to refuse on ethnic and traditional grounds. So the sections rapidly became home for many new families and all hope of maintaining a rational pattern of land use became impossible. Owen-Lewis had perforce to abandon the project. At the outset, the Land Use Team had forgotten an important precept: you must first study the people, before you evaluate the soils.[44]

Let us pause to consider Wright's "precept": "you must first study the people." This populist trope is iterated throughout Wright's texts – and it is not uncommon in contemporary development texts. The specific basis of Wright's populism was, to use a name from the social sciences, cultural ecology. Although he writes here that one should study the people *before* the soils, Wright's texts insist that the two go together: one studies the soils because they inform you of the culture, and the culture because it shapes (in its relationship with the soils) the potential development of the region.[45] Development discourse incorporates cultural ecology when it treats an evolutionary relationship between a culture and its environment as the basis of correct development knowledge. In Wright's texts on the Maya, this is a fundamental precept: *real* development is only possible when soils, race, and production are harmonized. Of course, this presupposes an organic, essentialist, trans-historical notion of race and environment: blood and soil. The notable effect of this trope for development discourse is that it solicits the development expert to the scene. The development expert is the unnamed second-person subject to whom Wright gives his advice: "*you* must first study the people."

Wright took his own advice assiduously. After writing *Land*, he spent much of his life studying the Maya farmer. His otherness (and Wright was only interested in Maya men) articulated that "desire and derision" typical of colonial discourse.[46] At one point in *Land* this articulation takes visual form. Because of the cost of including color photographs, Wright was allowed by the publishers to include only one photo in *Land*.[47] One might imagine that, since the text positions itself as a scientific soil survey, Wright would have selected a color photo of soil types. No. The photo Wright selected for the book portrays a Maya farmer fording a river (see figure 4.1). The photo's caption reads:

> Juan Sam is returning to his family in Otoxha after making a visit to Punta Gorda to purchase essential supplies – salt, soap, matches, kerosene and probably rum. You are seeing him on the last lap of his trip. He has paddled his dug-out canoe some 15 miles down the coast from Punta Gorda and then for 30 miles upstream on the Temax River. For the last nine miles he has carried his load overland and he has arrived at the main crossing of the Temax River – only three more miles to go. The deep, cool, green waters of the Temax are well stocked with machaca, tuba and other eating fish. At the point in the photograph, the river tumbles over a limestone ledge which forms a fairly safe crossing place – except in times of flood. To this ford comes an ancient Mayan trade route, linking the highlands of Guatemala with the shores of the Caribbean. Juan travels in company with the shades of his ancestors and he is probably aware of this as he hurries to reach home before night-fall.[48]

The photo and its caption reiterate the valences around the Maya farmer. We submit to empathize with Juan Sam's tiring journey, the long miles walked simply to purchase essentials. This empathy inspires our desire for development – to ease Sam's journey. Except that the photo and the text also work through a romantic Mayanist discourse: in Sam's movement through the "deep, cool, green waters" where they "tumble over a limestone ledge"; in the description of his honest labor;

Figure 4.1 The Kekchi Traveler (frontispiece to *Land in British Honduras,* 1959)

Source: Wright, ACS, et al. 1959. *Land in British Honduras* (London: HMSO).

and in his relation with the ancestors who share his journey. Wright celebrates the organic ties that bind Juan Sam, Maya farmer, to the land. And yet in a way that reminds us that his life is harder than ours, the reader's.

The photo serves as a way to consider how the ambivalences of colonial discourse frame Mayanist tropes of development. The photo is offered up as a representation of the real Juan Sam. And we read: "You are seeing him on the last lap of his trip." We are seeing the Maya farmer in a true representation of him. The realism of the photo, doubled in the performativity of the statement "you are seeing him," reflects modern metaphysics: to paraphrase Heidegger, the essence of modernity's epistemology is the representation of the real world as a picture.[49] And as Mitchell argues, European colonialism distributed the distinction between reality and its representation in and through its disciplinary practices.[50] These practices include the mundane act of Wright's selection of this photo for *Land*. By offering up the photo as a representation of the real Juan Sam, in speaking for him – reading his thoughts about his ancestors, no less – Wright *performs* trusteeship. Wright's invitation to gaze at "Juan Sam" as he trudges across the river (in the company of *our* shadows as they fall on the page) calls us to entrust ourselves in the care of this figure: the Maya farmer. In the space between the photo, the caption, and the reader, Western metaphysics and colonial discourse fold together and solicit development as a relationship of trusteeship and subalternity.

"Giving Civilization to these Indians": The Maya Indian Liaison Officer, 1953–1959

In the archive on Mayanism and development, *Land in British Honduras* marks a moment where colonial discourse about the Maya is becoming disciplinary. By framing the Maya farmer as a resource for colonial-national development, it foreshadows a movement towards the thematization of the "Maya community" as a space for development. In its mix of Mayanism, populism, and speaking for the Maya it reconstitutes colonialism as development trusteeship.

Land also played a crucial role in directing the institutional investments of the Belizean state. The first of these was the Maya Liaison Officer to whom we now turn. Earlier we read Wright's version of the story of the creation of the position of Maya Indian Liaison Officer. Other accounts cast doubt on the fact that the position was "Wright's idea." It should be evident that asking who came up with the idea is irrelevant. The ambiguity about the origins of the position is noteworthy, however, as it underscores its fundamental arbitrariness. Owen-Lewis recounted the events that led him to southern Belize in this way:

> I read that book. The book on Maya Indians [Morley's *The Ancient Maya*].[51] The book had a map in it. And the map had British Honduras on it. I started looking in Whitaker's Almanac or whatever to find out where British Honduras was, and who worked out here, and how to get a job out here, and I started writing.... Then I wrote to the Colonial Secretary who was number two to the Governor in British Honduras. I said, "what about a job? These are my qualifications." He said, "funny you should inquire, because a land-use survey has just done a survey there in Belize and a bunch of these villages which are very primitive, no

medical facilities, no schools, no access. We think somebody should be down there to see what he could do to help. Would you be interested?" And I said, "yes, it's exactly what I'm looking for."[52]

Owen-Lewis arrived in Belize a few weeks later, after a transatlantic trip on a banana boat. He received no training for his post. He arrived in Belize City, met briefly with the colonial officials, and left quickly to spend one night in Punta Gorda where he met with the District Commissioner before traveling up the Temax River to the remote Q'eqchi' village of Crique Sarco. His program was open-ended, with no particular goals to attain. His place within the colonial state bureaucracy was undefined and he commanded few resources. After leaving Belize City, he never received further instructions about his work.[53] In his words:

> The Maya Indian Liaison Officer. This was the job. I had no boss.... I didn't work for any department. I just – they put me there, and said, "get on with it." I said, "what do I do?" They said, "we don't know. It's your business. Get on with it. Giving civilization to these Indians. Help them any way you can." ... Initially, I had no money. The idea was that I had to find ideas, or whatever, and put them up to folks in a Government department, and they would supply me with the necessary cash to put it through. Like putting in an airstrip, getting a boat, cleaning up trails, getting a dispensary, whatever.

We should note that both times Owen-Lewis begins to explain exactly what he was sent to do, he says, "whatever." This is neither flippant speech nor parapraxis. It is exactly what he was sent to do: simply whatever he felt was needed to "give civilization to these Indians."

Owen-Lewis's work in Crique Sarco took two main forms. On one hand, he coordinated the labor of Q'eqchi' residents to build state institutions: a police station, a school, a farm demonstrator's house, and an airstrip. On the other hand, he coordinated development projects, such as the aforementioned land project. These projects were not determined through some planning process with the residents of Crique Sarco. In Owen-Lewis's words, the projects "just evolved naturally," naturally toward capitalism qua development. Their natural evolution was guided by the ways Owen-Lewis interpreted the task of "giving civilization." Initially he focused on learning Q'eqchi', supplying medicine to the sick, and learning about Maya agriculture. His initial intention was to change the Maya farm system by encouraging the planting of more commercial crops and livestock: "Planting coconut groves, planting oranges, planting cacao, bringing in cattle, the cattle business. Jack-of-all-trades." The commercial sale of pigs comprised one of these trades. The main limitation to pig export from the rural communities at that time was not the absence of a market, but the lack of transport to Punta Gorda. Owen-Lewis therefore bought a barge and put a motor on it to make regular runs to Punta Gorda with pigs:

> We used to have about fifty people and about a hundred hogs.... Whatever corn, or beans, or whatever they had to sell they'd stick it on the boat.... In the peak [we went] about every two weeks or something like that. It was quite good. It was really quite efficient. But then again, when I quit, that was it. The Government kind of pulled the plug on the whole operation. It lasted for, I think, about six months.

The boat fell into disuse and rotted after Owen-Lewis left Crique Sarco.[54] It is a fitting metaphor for the sustainability of Owen-Lewis's initiatives.

Consider the airstrip, for instance. One may ask whether clearing an airstrip was an important use of labor for a rural community that had never seen an airplane. It is unclear how Owen-Lewis motivated the men in Crique Sarco to contribute their labor to clear the airstrip from the forest, or why he felt it was a priority. The idea was clearly encouraged by the Governor, who wanted to be able to visit. (What was the point of having an official Liaison to the Maya if the Governor could not liaise?) The only visitors appear to have been the occasional government official: "There were the big-shots, the Governor used to fly in from time to time." When Owen-Lewis moved out of Crique Sarco, the reason to visit the area by plane left with him, and the planes stopped coming.[55] "The pilots didn't like it – thought it was damned dangerous. When I was there, I could twist their arms and get them to fly in, but when I quit there was nobody to twist their arms and they didn't fly in anymore." The other buildings built during Owen-Lewis's tenure were also abandoned.[56]

The thought that someone could be sent to rural Belize with the project of bringing development but without capital or specific plans appears shocking and ancient today. Today, were CARD or another development project to propose something similar for the same community, it would require drawing up a detailed plan, calculating potential costs and benefits, and (at least in theory) meeting with community representatives. This reflects the important shift that has taken place with the disciplining of development since the mid-1950s, when the Maya Indian Liaison Officer "job" was constituted. This distinct subject-position emerged in the ephemeral gap between the research for *Land* and its publication at a moment when colonial development discourse was still *becoming disciplinary*. The goal was explicitly development – yet this could still be defined as "doing whatever." This explains why the Liaison Officer needed no training: colonial development meant doing whatever any British colonial agent placed in that position felt should be done to develop the Maya farmer.[57]

Slash-and-Burn People

Owen-Lewis once described his position in Crique Sarco as "guide-philosopher and friend" to the Maya.[58] This begs the question: what philosophy guided his work? One answer can be found in *Land*, at the point where the Liaison Officer's role is explained:

> This officer will live in each of the main villages in turn gaining the confidence of the people, learning their problems and, where possible, showing the villagers how to overcome these problems. The main object in appointing this Officer was to try and advance the agricultural thinking of the Indians so they may play a full part in the agricultural development that must inevitably come in this part of the country.[59]

We should note that the tensions around the figure of Maya farmer noted earlier concentrate in the subject-position of "developer." The Liaison Officer is charged with "gaining the confidence" and "learning [the] problems" of the Mayas. This knowledge and psychological evaluation will allow the Officer to facilitate the "advance" in "agricultural thinking." The keen ambivalence around the *milpa* manifests in the argument for the power of the Officer. Is the Officer there to *liaise*, or to serve as *trustee*, of the development of the Maya? To ask this differently: is he there to speak *with*, or *for*, the Maya?

An answer can be inferred by considering the land scheme in Crique Sarco initiated by Wright and the Land Use Survey Team. It collapsed even before Owen-Lewis left, killed by its own success – if we accept his version of its history. Owen-Lewis explains the reason for the failure as a consequence of trying something that was "ahead of its time":

> Charles Wright had been working there . . . and he was way ahead of the game. He was very keen to find out if settled agriculture would work. . . . [I]ndividuals would be given a 50-acre block of land, and they were supposed to rotate it. And keep to their 50-acre block. There were

something like twenty 50-acre blocks, and this is all nice new land across the river....Relatives and friends of these lucky twenty would say to their brother or father or whatever, "can I have a piece of that for my plantation, too?" So we had about forty families living on twenty blocks. There was no way you could rotate it properly. It was just too, too far ahead of the game....We weren't ready for that sort of thing, not at that stage of the game....[T]hey had to clear it, but they were supposed to rotate it with cocoa, and cattle. It was way ahead of the game. Looking at it now, fifty years later, it was fifty years ahead of the game. Even now it would hardly work....[B]asically, the Q'eqchi' here are still shifting agriculturalists. They're slash-and-burn people.

We need not reiterate the findings of chapter 2 to interpret these remarks. The effect of the identification of the Maya qua "slash-and-burn people" should be clear. Here I would add only one further reference point, to emphasize how Owen-Lewis's subject-position underscores the territorial quality of capitalism qua development. We noted earlier that Owen-Lewis's desire to travel to Belize and work with the Maya was inspired by reading Morley's 1946 text, *The Ancient Maya*. Morley's text was one of the first attempts to write an accessible assessment of the Maya that adopted the style of scientific, modern archaeology, drawing from the vast literature that emerged in the early twentieth century as many of the ancient Maya cities were scrutinized by the second generation of Mayanists. For Morley, the Maya farm system was the essence of "Maya" culture: "the most basic fact about the Maya civilization," Morley explains, is that "their culture was based directly upon, and derives straight from, agriculture as applied to the cultivation of corn."[60] Before traveling to Crique Sarco, Owen-Lewis would have read: "the modern Maya method of raising maize is the same as it has been for the past three thousand years or more – a simple process of felling the forest, of burning the dried trees and bush, of planting, and of changing the location of the cornfields every few years."[61] Here the Mayanist narrative – with its temporality that presupposes that today's "fallen Maya" can be redeemed through a movement that parallels the West's development – is implicated in Owen-Lewis's judgment that the land scheme was "way ahead of the game."

Colonial spatiality is also implicated in this game. Consider figure 4.2, a "physiographic map of the Maya area" taken from Morley's 1946 text. This is the map, we will remember, that inspired Owen-Lewis to leave England ("The book had a map in it. And the map had British

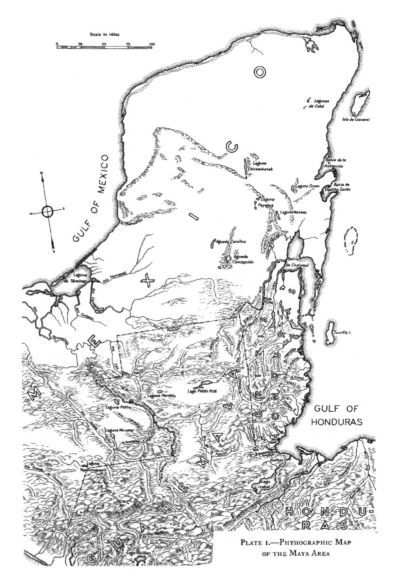

Figure 4.2 "Physiographic map of the Maya area" from Morley, 1946

Source: Morley, Brainerd and Sharer – *The Ancient Maya*, 3rd edition. © 1946, 1947, 1956, 1983, 1994 by the Board of Trustees of the Leland Stanford Junior University.

Honduras on it. I started looking in Whitaker's Almanac . . . to find out where British Honduras was, and who worked out here, and how to get a job"). *Here*, the map says, is "the Maya area." Like most modern maps, the mode of representation is realist: there is a North arrow, a scale, and the physical features of the landscape are drawn in three-dimensional form. Reading this map, we believe that we could use it to move through this space, and that we are looking at a good or true representation of the Maya world. *There*, Owen-Lewis might have said to himself, is an accurate representation of the lands in which the fascinating Maya live. And yet, although it is a map of "the *Maya* area," the space of the map is divided into the territory of four nation-states. Only on this basis could Owen-Lewis discover that "British Honduras" happened to be in the middle of the Maya area and repeat, metaphorically, the map's spatial performance: in his desire to enter Maya lands, he discovered that he was already affiliated with them, within reach of them, in England. His subsequent superimposition upon Crique Sarco could not but contribute to the territorialization of those Maya lands as British.

Four hundred years after Las Casas (see chapter 3), here we can discern the reason of territorialization: the modern map, a representation which is said to stand in for the real, produces a space said to be "Maya" – without specifying what constitutes this "Maya" – *only to overlay the Maya land within non-Maya territory*. Until the text opened this particular space before Owen-Lewis, there was no *there* there. Although the map presents itself in a modern, realist mode ("correct" orientation, scale, etc.), it achieves its effect – that of producing a "Maya land" superimposed by territories – through the extension of the particularly Western metaphysics that treats the world as divided: real versus representation, world versus map.[62]

Reading the Liaison Officer Position

Those who have commented on Owen-Lewis's accomplishments in Crique Sarco have given sympathetic readings. For instance, when Wilk writes that "Owen-Lewis began a number of successful projects in Crique Sarco, including construction of an airstrip, a regular boat service, and communal cattle pastures, but it was wishful thinking to send a single man, however talented, to develop an entire region," he implies that the British should have sent more of their talented Liaison Officers.[63] Frederick Nuñez, who evaluated Owen-Lewis's work in the

1970s, noted an apparent contradiction at the heart of the Liaison Officer's subject-position:

> The Liaison Officer had got the opportunity to work with [the Maya] because he had grown interested in their culture. This interest was two-fold, consisting partly of a positive attraction to Indian culture and in part to an increasing distaste for Western society. It is surprisingly ironic that this was the applicant who was cast in a role essentially designed to Westernize the Kekchis.[64]

Nuñez therefore finds it ironic that Owen-Lewis, who loved the Mayas and felt distaste towards the West, was sent to develop the Maya. This is an understandable but limited reading, since as we have seen, the agent of development is *supposed* to embrace the Maya farmer. Furthermore, Nuñez presupposes that Owen-Lewis was attracted to the Mayas but not the West: two valences, not in conflict, set against the fact that he had a job to do. Yet Nuñez leaves out one crucial valence, shared both by the colonial state and Owen-Lewis: the pull of reasoning that *the Maya need to be developed*. There is no irony that the person who was sent to act as the agent of development felt a "distaste for Western society." Such distaste has motivated development in its historical affiliation with capitalism, for as we saw earlier, development supplements capitalism as a rectifying discourse. And as Wright insists, developing the Maya farmer *requires* a "positive attraction to Indian culture." There is no contradiction between an "interest" in an indigenous culture and a desire to guide it through the process of development.

Ironically, Nuñez's text, which is ostensibly a critique of Owen-Lewis's work, reproduces the colonial disciplinarity manifest by Owen-Lewis's approach to development. Nuñez concludes that Owen-Lewis failed because he had insufficient knowledge of the Maya culture:

> In all of its aims it is evident that Government's intention was to act in the interest of the Indians and in the interest of the development of the region.... But they were unable to do so because of their complete ignorance of the very people whom they set out to help. Changing the milpa system for fixed-plot farming was not a simple technical [problem].... The Government never attempted to understand Kekchi culture.... *Had Government simply gathered this basic data, their plans might have been far more effective.*[65]

Here we must remember that it was the desire to "gather... basic data" about the Maya that led Wright et al. to call forth the Liaison Officer.

In the anthropological view, the failure of development intensifies the desire for data on the Maya. Thus, the argument for development – on the basis of disciplined knowledge of the Maya farmer that one must transform – reiterates itself while soliciting "Maya" culture as *data*, i.e., as a resource for development.[66]

If *Land in British Honduras* marks a turning point in colonial discourse, then the creation of the Maya Liaison Officer position was its first institutional effect. The position was created to protect the colonial state from liberal criticism in Britain from advocates of indigenous welfare. Its creation and implementation was intended partly to assuage these criticisms, which were shallow and entirely amenable to – indeed, solicitous of – colonial trusteeship. The solution was the subject-position of Maya Indian Liaison Officer. The results of his work, measured in terms of changes in the material conditions of life in Crique Sarco, were basically nil. But even development "failures" produce effects. Two in particular have outlasted the boat, the airstrip, and the 50-acre scheme in southern Belize: the identification of the *milpa* farmer qua racialized "Maya" and the solicitation of development-as-trusteeship by colonialism. These effects were produced by the discourse on Maya farming, including *Land*, and not the Maya Indian Liaison Officer alone. But they are all that remain from his tenure.[67]

The Conflict over the Columbia River Forest Reserve

If the first material effect of *Land* was the position of Liaison Officer, the second was something "natural": a forest. In 1953, during his soil survey of the forested headwaters of the Columbia River, Wright noticed something unusual about the landscape: although the tree canopy was among the tallest in Belize, there was almost no surface water.

> A subsequent study of climatic data suggested that this forest probably belonged to the category of humid wet tropical rainforest... that once fully established is capable of controlling its own internal moisture and many other environmental conditions. [...I]f so, then it was the most northerly example of this category of forest so far identified in Central America. Clearly, this forest warranted further study in the future, and thus should be protected from exploitation.[68]

Wright saw a forest that had not been recognized as such before. Thus the Columbia River forest, a socio-natural space that had been used and reshaped by human labor for thousands of years, became rearticulated and thematized in a new way. With Wright's assistance, it became an object for the attention of the colonial state:

> I made sure that this sector of the Toledo landscape was removed by the colonial government from the general category of Crown Land. The only alternative available at that time...was to persuade the Chief Colonial Forester into gazetting the area as a forest reserve, and so it has survived to this day.[69]

As a direct result of Wright's lobbying, in 1958 the colonial government gazetted 132,750 acres of wet tropical forest in the northwest corner of the Toledo District as the Columbia River Forest Reserve (CRFR). Under regulation by the Forest Department, the CRFR was created to preserve "mahogany forest from clearing for cultivation and to secure land for government re-forestation with mahogany and other species."[70] No provision was made during the creation of the CRFR for the many uses of the forest by the Maya communities that surround it on three sides.

As a result of its designation as a Reserve, almost no logging took place in the Columbia River forest for 35 years, apart from modest timber collection by local Mayas. But in the early 1990s, with the state under pressure by the IMF to increase exports, the Ministry of Natural Resources granted a series of logging concessions for the CRFR.[71] The new and complex logging concessions were to be regulated by Belize's Forest Department, an institution with very little institutional capacity.[72] The Department is chronically underinvested and understaffed,[73] and therefore forest licenses are allocated in a "largely ad hoc, piecemeal manner"[74] and royalties are collected inefficiently or not at all.[75] To compound these problems, in December 1995 an IMF-encouraged "retrenchment" of public employees swept Belize. The Forest Department was hit especially hard, losing a quarter of the staff, including some of its most experienced employees.[76]

These problems were explored in the late 1980s by a group of foresters from Britain, the US, and Belize through a series of meetings that led to the creation of a "Tropical forestry action plan" for Belize. This action plan proposed a set of reforms aimed at producing a sustainable management approach to Belize's forests with foreign assistance from the US and Britain.[77] In 1993, the British Overseas Development Agency created the Forest Planning and Management Project (FPMP) to provide technical expertise for rational forest management. FPMP's first major project was to write a management plan for the Columbia River Forest

Reserve which stipulated 83 conditions concerning road construction and tax payments, as well as area and volume controls for logging.[78] FPMP felt that the CRFR should be exploited on a rotating 40-year cycle (based on the calculation that valuable hardwoods could regenerate in 40 years); the CRFR was therefore divided into forty 1,200-acre blocks. The licensee was to be allowed to operate in only one block per year and to cut no more than 300,000 ft^3 of timber. The plan called for extensive Forest Department oversight and the involvement of the neighboring Maya communities.[79]

In February, 1995, "a Belizean registered company with Malaysian financing"[80] called Atlantic Industries, Ltd. (hereafter AI) submitted an application for the license to log the CRFR. The FPMP reviewed AI's proposal for the CRFR and found that it was "a very superficial document [that] clearly lacks professional preparation and analysis[,] not something one would expect to see from an investor who is preparing to invest $3.5 million in a timber operation."[81] Nevertheless, in April, 1995, the Minister of Natural Resources granted Forest License 6/95 to AI. AI's work was bound by the terms of the CRFR management plan – but the company ignored these from the beginning. Crucially, the plan called for the licensee to begin logging in one particular place (north of the village of San Pedro Columbia) and then haul logs to a sawmill outside of the CRFR. But in September, 1995, the company moved into another area of the CRFR (northeast of the village of San Jose) and cleared a few acres to build their sawmill. From this mill site they began cutting new logging roads throughout the CRFR. Leaders from San Jose spoke out against these actions immediately. San Jose's Chairman obtained a copy of FPMP's plan and traveled to a national environmental conference to enumerate AI's breaches of the contract: cutting in the wrong area; harvesting trees without Forest Department clearance; and failing to consult with the community.

Serendipitously, this coincided with another set of events that catalyzed a new Maya movement. In early 1995, a schoolteacher named Julian Cho was elected Chairman of a new group created by the alcaldes to study the legal status of Indian reservations – the Reservation Lands Committee (RLC). As Chairman of the RLC, Cho organized a research program and held weekend meetings in different Maya villages to discuss land rights and raise awareness of the risks facing Maya communities.[82] Cho wrote to the national newspapers about AI's logging in the CRFR.[83] Cho used the controversy surrounding the CRFR as his illustration of the need for the Mayas to struggle to win secure collective land tenure. He articulated a similar message in a series of newspaper articles that argued that the future of Maya culture and the forests were intertwined – around the question of indigenous land tenure:

> The local Mayas, who see themselves as the custodians of the rainforest, regard [the logging concessions] as blatantly disrespectful to their dignity. Any foreign intervention is questioned within the communities which border the forest reserve and by all of the Mayan communities within the District. Given the tenuous character of Mayan land claims in the region, an intrusion of exploitative, land-destroying operations directly threatens the Mayan communities.... Mayan people in Toledo need much stronger legal land tenure to existing reservations and ... to consider the proposition of a Maya homeland.[84]

By asserting the indigenous and social character of the CRFR, indigenous land claims were articulated with the environmental arguments against the logging concessions.

The state responded to these claims through a series of press releases, radio announcements, and political speeches that framed the conflict over the forest in two important respects. First, the state framed the debate over logging concessions as a merely technical (and therefore apolitical) question of proper environmental management. The experts of the FPMP and the Forest Department justified the concession on the grounds that the Management Plan protected local communities around the CRFR[85] and argued that Maya farmers, not logging companies, constituted the major threat to the forests.[86] A British forester advising the government dismissed Cho's claims that the logging would damage the forest, claiming that Cho "exaggerated environmental concerns as a proxy argument for the real issues which concern the local Maya people, namely land ownership and who benefits from timber exploitation. Both of these are political concerns."[87] What is notable here is not the claim that Mayas like Cho were more interested in questions of landownership and the just distribution of benefits, but the assumption that such *political* concerns were beyond the purview of the *technical* experts – and not the grounds upon which to judge the fate of the logging concessions.

The government's second move was to frame the conflict at a broader economic and territorial *scale*. Through press conferences advertising the importance of forestry in the *national* economy, the state emphasized that all Belizeans stood to benefit from AI's investments in the CRFR. The effect of this discourse was to suggest that the opposition was purely local, thus marginal. Yet the state felt obliged to claim that these communities had been consulted properly about the concession: the Forest Department thus claimed that "several meetings and consultations were held with villagers . . . prior to the development of the forest management plan for the area."[88] Forest Department internal reports show, however, that only one meeting was held to "liaise" with the Maya communities and one FPMP social forester noted: "We should not have assumed assent to the plan based on one meeting with a village."[89]

As the debate raged, AI continued logging and tensions grew. On December 2, 1995, Julian Cho led a march of hundreds of Maya farmers through the streets of Punta Gorda to denounce the "Malaysians" in their forests (see figure 4.3). To quell the growing storm, the Forest Department sent a team to Toledo to lead a group of Maya leaders with AI's managers for a hike through the CRFR.[90] The foresters were met by an angry but disciplined group of hundreds of Maya activists, who blocked their entry into the CRFR. The group only allowed the hike to go forward after the foresters promised that they would report back to the group (with Julian Cho and other Maya leaders) immediately upon returning. AI managers looked tense as they entered the forest with the Maya leaders and the Forestry officials, and soon revealed that they had never read FPMP's Management Plan.[91] This fact was promptly publicized by the Maya leaders, who then prepared their own version of a Management Plan for the "the Columbia River Forest Reserve *and Indian Lands*" with a set of strict conditions to be met before allowing further logging; one condition read: "Whereas the Mayan people are the original inhabitants of this territory that is currently controlled by the State, and this territory is the place where we find our origin, being, philosophy, science, and the roots of our languages, Mopan and Q'eqchi', we have agreed to claim a territory for our Mayan Nations."[92] The Mayas' decision to frame the struggle over the Columbia River forest as a territorial conflict between two nations marked a decisive shift in the movement's representation of their aims.[93]

Confronted by an emboldened Maya movement, the state was compelled to close AI's sawmill near San Jose. The Maya thus succeeded in temporarily forcing AI out of the CRFR. But only a month later, a sympathetic official in the Forest Department leaked a sketch map to Julian Cho that revealed that the state had conceded nearly the entire historical area of Maya land-use in southern Belize for logging; one new concession to AI covered more than 100,000 acres alone. AI had quietly built a large new sawmill a few miles to the east. Reiterating the colonial accumulation project to the letter, the company announced plans to export unfinished lumber to the US market.

Cho and the Maya responded by strategically and spatially broadening their movement. In April, 1996, their Chairmen, Julian Cho and Santiago Coh, traveled to Washington, DC to mount an international

Figure 4.3 Julian Cho (at left) leading march through Punta Gorda, 1995

Source: Photo taken December, 1995, by an unknown photographer. The photo was given to Pio Coc, one of the leaders of the march, who gave it to the author. Efforts to determine the photographer have been fruitless.

media campaign,[94] raise money, and lobby for pressure on the Belizean state.[95] Upon returning to Belize, the Toledo Maya Cultural Council and the Alcaldes Association announced plans for two major projects: to map all of the Maya lands in southern Belize, and to launch a lawsuit against the government for infringing upon their indigenous land rights.[96] I discuss these still-unfolding projects in subsequent chapters. For here we must return to Charles Wright.

Charles Wright and the Social Nature of the Columbia River Forest

For a tropical forest, the CRFR has an intensely social nature.[97] This forest reserve became an unlikely meeting place for development discourse, ecological science, multinational capital, and a renewed indigenous land rights movement. And at the center of this ensemble, literally organizing the complex of texts, facts, and political interventions, was Charles Wright. Wright was implicated in the conflict over the CRFR in many ways: as the scientist who originally advocated for recognition of this forest as an ecological space with particular qualities; as the colonial official who pushed for the creation of the Forest Reserve; and as a partner of Julian Cho in the struggle against the logging. Only months before Atlantic Industries entered the CRFR, Wright took up residence in the back of the new office of the Belize Center for Environmental Studies, an environmental NGO not inclined towards confrontational politics. Wright was on familiar terrain, working in a sparse room in a wooden, colonial-era home in Punta Gorda – just a block down Front Street from the former colonial District Headquarters. With patience and charisma, he forged this space into a salon where Mayas and environmentalists gathered to debate their strategies and concepts. Wright even became one of Julian Cho's most trusted interlocutors: between October 1995 and January 1998, Cho and Wright met in his office more than fifty times to discuss this social forest and the politics surrounding it.[98]

People gathered around Wright because of his pen. Although Cho was undoubtedly the leader of the Maya resistance to the logging concessions, Wright was the anti-logging movement's most prolific critic. In a series of commentaries written between 1995 and 1998, Wright savagely criticized the state's forest policies and the treatment of the Maya.[99] These texts are notable for their proximity to the very themes and questions that dominate his studies from the colonial period, including *Land*: land use, the Mayas, and sustainable development.

Wright's late writings reveal shifts in his thinking, in at least two respects. The first concerns the value of outside experts – like Wright himself of four decades earlier – for the development of indigenous communities. Consider Wright's argument in an important essay from 1995, entitled "A history of the Columbia River Forest Reserve."[100] In this text, Wright offers a social history of the forest that calls into question the very discipline of development that he helped to bring into being: "Every few years a new team of international agronomists arrives in Toledo to show the Mayas how to increase production of cash crops on a sustainable basis, but so far little of anything helpful has resulted."[101] Coming from a colonial soil scientist who came to Belize explicitly to "show the Mayas how to increase production of cash crops," this claim is at once poignant and ironic. Not only did he disabuse the practice of relying on foreign experts – including the British foresters of FPMP – who came to speak for the Maya about develop-ment; after decades of studying the soils and farming in southern Belize, Wright explains, he had come to conclude that "the traditional Maya method [produces] the best crops."[102] Development therefore must be thought on the basis of the Maya farm system.

Second, Wright's writings about the CRFR reflect a radicalization of his earlier conception of the appropriate involvement of local com-munities in development. His most acerbic writing from this period criticizes the state's refusal to recognize indigenous rights to shape the development of the lands. His earlier colonial populism had been sharpened into hard demands for state accountability and genuine participation of Maya communities in development planning:

> Little or no attempt [to consult with Maya communities] was made by Government, by the Minister in charge of forests, or even by local forestry staff. The Draft Management Plan [for the CRFR] was simply slapped down in front of the villagers as something they would have to learn to get along with. Considering that the area involved was precisely that where many Mayans still had lingering hopes for creating their homeland, the method of "promoting local involvement" almost amounted to delib-erately provoking the Mayas to reject the scheme out-of-hand.[103]

But as much as Wright's epistemic habits evolved, certain colonial modes of thinking and writing were never overcome. The first and most important was to speak *for* the Maya, to discern what was good for them and to report it to the public. Wright never hesitated to assume the old, confident position where he would simply state what it should take to get the Maya settled down again and the forest protected. Consider his argument from the same essay:

> The solution to this rapidly heating-up new scandal is clearly to put the whole scheme on hold until the Forest Department can find enough money to recruit and train the extra staff needed to police the forest-culling activities called for by the scheme; and to give the Mayan Indians time to study the many good points in the pioneer Forest Planning and Manage-ment Project. Malaysian collaborators should be quietly eliminated.[104]

Here is colonial politics *redux*: bring in more police; give the Mayas an education in what is good for them; and "quietly" separate racial groups.[105]

If this tendency to speak for the Maya revealed the vestige of colonial epistemic habits, its most pernicious effect could be felt in his reflections on population and the limits to Maya land rights. As much as Wright advocated for the concept of "Maya lands," he never ceased to reiterate the Malthusian arguments he made forty years earlier:

> The Maya population on Indian Reservations has long been busting at the seams. The Maya have increased greatly in numbers in the past seventy years and are wistfully looking at adjoining forested land as a logical place to which they could extend their traditional milpa farming for the production of subsistence crops.... Some Indians prefer to dwell on the possibility of amalgamating their reservations to create a large "homeland" as has been done elsewhere to preserve a particular kind of eth[n]ic way of life. Such a solution may certainly help to preserve ethnic customs for a short while, but would be doomed to eventual failure as their population inevitably increases; and in Belize there is no unused agricultural land to accommodate the repeated need for an enlargement of such a homeland.[106]

The indigenous claim to justice and the critique of colonial territorialization is thus rejected by the hard ecological facts in Wright's firm grasp. There is no space for a Maya homeland. The confidence of the father of the Columbia River Forest Reserve thus settles the Maya once again.

Conclusion

> The Maya of today must be amongst the most studied race on Earth, to judge from the enormous number of published and unpublished reports by sociologists, socio-economists, socio-anthropologists, and many more (including myself – an amateur socio-apologist). We all make the mistake that we expect to be able to identify the cultural characteristics that give the Mayan race a special individuality.
>
> A. C. S. Wright (1995)

> I personally am becoming reconciled to a certain loss of biosphere research which must inevitably result from commercial development of the Columbia Forest Reserve, perhaps hoping that not all will be lost if a real sustained natural timber production scheme can be effectively managed. But to wait almost fifty years only to find that a process of environmental rape has been contrived for the ultimate benefit of a few non-Belizeans (helped by a small group of Belizean nationals), then I certainly cannot die a happy person. I feel that I have, in some measure, let down my many Mayan friends. *Ulahapchik.*
>
> A. C. S. Wright (1995)

In 1997, Charles Wright died in Punta Gorda. He was 83 years old. His passing was noted mainly by Wright's Maya friends, who took him from his rented hovel on Front Street to the public hospital, cared for him in his last days, and laid him to rest.

What are the lessons to be learned from reading the texts Wright produced as interventions into the debate over the fate of the Columbia River forest? What can we learn from the attempts of a colonial soil scientist who spoke out in defense of Maya-centered development, against abuses of state power, and chastised the foresters who worked to erase the question of indigenous justice from the conflict over the CRFR?[107]

The first lesson is that one does not simply overcome or step away from colonial discourse. As we have seen, Wright never lost his Malthusian tendencies to reduce questions about Maya land use to population; nor did his tone ever betray a desire to cease speaking for the Mayas. This could be regarded as a personal limitation, but a reading that treats this as Wright's problem cannot explain its pervasiveness or its power.[108] Instead, I suggest that we read the limits of Wright's texts as effects of the discourse about the Maya and development.[109] The limits of his approach cannot be understood apart from, or separated from, the archaeology of the colonial discourse on Maya farming.

But we should also recognize the shift that occurs in the postwar discourse of development. The second lesson is that colonialism solicited a subject-position for those it would develop. In southern Belize, Wright's texts were part of an archive that thematized the "Maya" as a racial and cultural category. In so doing, the "Maya" and "their farm system" became objects that exhibited the peculiar quality particular of many "others" in colonial discourse: they are figured as essential to Mayaness, but in need of fundamental reform; as cultural resources for development, yet in need of trusteeship; and so on. This explains the ambivalences in Wright's texts that concentrate around the articulation of Maya difference.[110]

This makes Wright's comments on Maya identity, quoted in the epigraph, more compelling.[111] Wright is correct when he says that "the Maya of today must be amongst the most studied race on Earth" – although we should bracket the assumption that the Maya comprise a race – as a check of the anthropological literature will reveal. Wright poses this as a problem, and includes himself (in a self-deprecating and self-reflective gesture) in the list of experts that have produced this Mayanist archive: "sociologists, socio-economists, socio-anthropologists, and...myself – an amateur socio-apologist." What explains Wright's desire to see himself as an "amateur socio-apologist"? For whom does Wright apologize? To whom is the apology directed? Let us hold these questions for one moment to see where the text moves. Wright then begins to call into question the essentialism that has under-pinned the Mayanist literature: "we all make the mistake that we expect to be able to identify the cultural characteristics that give the Mayan race a special individuality." This is true. Wright has put his stylus – the one that so desired to pin down the wandering Maya farmer – directly on the essence of Mayanism. This amateur remark, this dot in the mass of the archive, marks the farthest point towards a critique of Mayanism in Wright's texts.

And as though he senses the waters are getting too deep, Wright immediately turns back to the shore provided by archaeology's firm ground: "We might have had more success [in identifying Maya cultural characteristics] if we had started our inquiries some time 2,000 years ago. The Maya of today has very tenuous links with the distant past, and his instinctive reactions are often more related to the not-so-distant past." The Maya have changed. They are not like they used to be – their links to the past are "tenuous"; they were more fully themselves in the past. "He" is still "Maya," yes, yet anachronically so: he is in the right space, but in the wrong time. And still his "instinctive reactions" are not rooted in the deep time of the ancients, but rather in "the not-so-distant past." Perhaps that is why Wright concludes that today the "Mayas are not very much different from ourselves." They, like us, are stuck in the now, no longer ancient, no longer tied to the land and bound to ancient rhythms. Such nostalgia for an imagined past, cloaked in humanism, pervades the colonial archive.

This leads us to the third lesson. It is not contradictory that Wright participated in both the colonization and development of southern Belize, and also that Wright began in his later texts to clear space for movement towards rethinking development. *Land* designates a crucial space in this archaeology, where the Maya and their farming system are transformed from a mere object of colonial reform into a resource for development. This is not simply the local application of a global process – the post-World War II "construction of development." It is, rather, a creative intervention that responds to the solicitations of colo-nial power, power that calls for a disciplinary form appropriate to colonialism qua development. The culture of the Maya farmer is thus identified as a resource for development knowledge.[112] This form of knowledge is constituted as a discipline that binds colonial power with development as a form of trusteeship. Remember that *Land in British Honduras* was created as a result of research conducted to respond to concerns about Maya welfare. In a parallel way, the "resolution" of the Maya claims to the Columbia forest was to create the aforementioned agricultural development project called CARD.[113] Part of the power of development lies in the bond it forms between resource management, economic development, and state territorialization. There is no ambivalence here – only concerted, directed power: the claim to space by the capitalist state.

Any attempt to try to understand and intervene into struggles over nature in the Americas requires a serious engagement with the facts of colonialism. There is no space on these continents that can be understood outside of, prior to, or apart from the facts of the colonial experience. The living legacies of colonialism compel us to try to account for the ways that colonial knowledges are always already at work in thinking about environmental conflicts. Between *Land in British Honduras*, the many consultancies of the later years, and his writings on the conflict over the Columbia River forest, we must ask: why is there never an attempt by Wright to situate the analysis of the conflict, or the indigenous claims against a colonizing power, in terms that account for the "objective" knowledge and power of the one called to do the analysis? Why is Wright, along with the Mayanists, ecologists, journalists, and kindred development consultants unable to see their complicity in the reproduction of these relations of power and knowledge that have, in a strong historical sense, produced the very antagonisms and conflicts they purport to empirically describe? Or, to borrow a question from Edward Said, posed about George Orwell's writings on the working class in England: "Is there to be no remarking of the power that put the reporter or analyst there in the first place and made it possible to represent the world as a function of comfortable concern?"[114]

Notes

1   This correlates with Homi Bhabha's insistence that colonial discourses and texts are "ambivalent at the origin," i.e., that texts crucial to the colonial project – Bibles, reports, textbooks, etc. – contain oppositions and parapraxis that can be interpreted as valences reflecting tensions within colonialism.

2   Wright et al. (1959). Wright is the lead author of *Land in British Honduras* – he is credited as such in the text, and his distinctive writing voice is clear throughout. However, the final *edit* of the text was carried out by his colleague, D. H. Romney, leading many to cite the text as "Romney et al."

without noting that Romney was editor, not lead author. To further complicate matters, Wright himself often cited this text as "Wright 1958," since (I imagine) he had a copy of the draft in 1958. I cite the text as "Wright et al. 1959," since *Land* is not an edited volume, but a book mainly written by Wright; 1959 is the date of publication in London.

3   Wright (1996a, 1996b). The latter is also useful for reconstructing the history of interventions since the 1950s and also for understanding Wright's later views on Maya agriculture.

4   Escobar (1994).

5   By "polyvalent" I mean motivated by multiple distinct and conflicting tendencies; I use this term in contradistinction to the now-common postcolonial notion "ambivalent" (see note 1). The complexity of Wright's multiple investments in historical, environmental, indigenous, and development efforts was especially pronounced in his last years, when I knew him: in one two-year period, he worked as a consultant for the Belize Center for Environmental Studies, a conservation NGO; wrote an important report for the Indian Law Resource Center, an indigenous rights NGO; consulted for IFAD on a development project; and served as confident to Julian Cho. He did such work at the end of his life out of commitment to these overlapping-but-distinct causes. He lived ascetically and spent his meager wages on rum, cigarettes, and meat for his dogs.

6   Said (2002 [1995]: 463).

7   His father was a university lecturer in physics who served as Director of Naval Scientific Research during World War I, for which he was knighted.

8   When he arrived in Belize in August, 1952, Wright was 37 years old and brimming with confidence from his successes in New Zealand and the war. This confidence permeates his work and texts of the ensuing period, which – notwithstanding subsequent surveys for the FAO that took him to more than thirty countries in the next four decades – consistently focused on southern Belize. Most of Wright's late works are unpublished. They are in the care of the Archives of Belize. I thank Luis Avila and Will Jones for their assistance in working through the Wright archival collection.

9   Stevenson, N. 1933. "Indian Occupation of the Toledo District." AB, MP 266–33. See also: Anon., 1934. "Indians in the Toledo District." PRO CO 123/348/9; Anon., 1937. "Report of the Forest Department British Honduras for the year 1936." AB, Annual Reports of the Forest Department; Anon., 1948. "Maya Indians." PRO CO 123/397/6; and (4) Anon., 1953. "Maya Agricultural Practices." AB, MP 47–53. These texts reflect the increasing density of the discourse about Maya farming in this period.

10 Ibid. For the ICMW the welfare of the Maya was a function of settlement, education, and capitalism qua development. The ICMW proscribed a specific curriculum for the Maya that included the geography of the colony "with only incidental facts regarding other places." History would be "taught incidentally with *Geography* – simple stories about the Colony

and its connection with Britain and the Empire, *aiming to make people Colony- and Empire-conscious*" (ICMW, 1941. AB, MC 1819: Appendix I, my italics). This report stimulated a new round of debate and research, including a report by the Social Welfare Officer in the West Indies that proclaimed, "The Mayas are dying out at a frightening pace" (letter from Ibberson to Macintosh, August 19, 1946. PRO CO 123 397/5). Ibberson reported the following "disturbing things" found among the Mayas of Belize:

1. There is no prohibition of selling liquor.... The Mayas get drunk and howl like dogs. In the South the women drink as heavily as the men.
2. Their infant nutrition is abominable.
3. [S]anitation...is nearly non-existent except for the services of pigs & dogs....
4. There is no infant welfare service. There are no maternity beds....
5. The Mayas are dying out at a frightening pace. Some figures collected by St. Aubyn were sensational.
6. San Antonio, in the South, has a group of Mayas from Guatemala who are the worst. The Jesuit missionary recently sent there is approved & made the senior boy of the school recite the numbers who were born & who died in the village last year in an address to me. The Jesuits are from USA & work with money from USA...but can't tackle the women which is essential.
7. The traditional crafts are stone dead, in painful contrast to Guatemala where they are vigorously & successfully exploited & the products poured into [Belize].

What is astonishing about Ibberson's list is that her claim that "The Mayas are dying out at a frightening pace" is listed somewhere down on the list from alcohol abuse and before the death of traditional crafts. Her fear is that American Jesuits will report on the mistreatment of the Mayas and thereby facilitate US "anticolonial" pressure. The letter concludes with this telling remark: "Small as the problem is, I think it might blow up into a scandal."

11 Letter from G. F. Seel to F. C. Hawkesworth, March 13, 1947. PRO CO 123 397/5. The Social Development Department was created in 1949 with a Colonial Development and Welfare grant. The department was comprised of "the Social Development Officer, two Assistants, three Co-operative Officers, a Handicrafts Instructor and a subordinate staff" (Anon., 1951. "British Honduras Development Plan – Social Development, p. 1. AB, MP 311–G). The only reference in this text to the Toledo District comes under the heading "Maya Welfare": "The strengthening of the Agricultural Department and the resuscitation of the Toledo Agricultural Station will enable a trained Agricultural Officer to devote more time to the special problems of the Maya Indians during the next four years, and the necessity to appoint an administrative officer with special *anthropological training* to take charge of their affairs" (p. 4, my italics).

12   The CD&W Act was passed in 1940 after a decade of criticism of colonial policy, coupled with anticolonial pressure abroad, seriously threatened the stability of London's hegemony over the empire. The Act consolidated the new colonial framework of "constructive trusteeship" and aimed at "restabilizing the empire and diffusing criticism of British colonial rule" by providing "a new justification which would legitimize the perpetuation of colonial rule" (Constantine 1984: 259). A meticulous narrative of the research for *Land* is chronicled in Wright's diary, which was carried by one of Wright's Maya friends to the Archives of Belize after Wright's passing.

13   The major finding of this section is that "The country is not well placed geographically" (Wright et al. 1959: 4).

14   Ibid., p. 7.

15   This strikes me as quaint, for agronomy and soil science have long ago been replaced by economics as the hegemonic discipline in development studies. Today the development experts would not write a development plan for Belize by sending a team of soil scientists with a blank check to survey natural resources. We should acknowledge, without nostalgia, that the soil scientists strayed farther from the capital, talked with more peasants, and looked at the earth.

16   Wright et al. (1959: 39).

17   Ibid, p. 39.

18   Ibid., p. 261.

19   Ibid., p. 124.

20   The only published criticism of Wright that I have found is that of Wilk, who argues that "instead of trying to 'pin down the wandering milpa' so that economic development can take place, it would be better to promote rural production and provide village services first – then the milpa farmers will settle down by themselves" (Wilk 1997: 5). In fairness to Wright, his texts from the 1990s reflect a change of tone – in agreement with Wilk.

21   Wright, et al. (1959: 281).

22   Ibid., pp. 281, 124.

23   Ibid., pp. 128, 132.

24   Ibid., pp. 124, 128, 132, 281. Wright et al. write that the Q'eqchi' is "a willing and able farmer but he is a poor mathematician" (p. 128); as for the Mopan, their "material advance" is "outstripping the development of their sense of values, and on a sudden impulse they may throw away the labour of weeks on something which has no use or meaning in their home life" (p. 38). In other words, this colonial resource calls for *economy*.

25   Ibid., p. 38.

26   Not only the Maya, but all of Belizean society is described in terms of race – with no mention of gender, class, or other elements that constitute identities. Ibid., pp. 33–47 ("The pattern of the people and their livelihood").

27   Ibid., pp. 35–9.
28   "The Maya Indian *returned* to British Honduras from about 1850 onwards" (ibid., p. 37, my italics). The ancient Maya are discussed in the first half of the section on the "historical summary of land use" (pp. 110–15). A map overlays the ancient Maya sites with the free-draining black soils on limestone to suggest that the ancient Maya preferred those soils (see figure X of *Land*, "British Honduras Ancient Maya Sites"). Wright claims that "the link between the ancient Maya Indian and the various Mayan racial groups... in our time is a tenuous one" (p. 113). This language is reiterated by Wilk: "Though it is common to make some connection between the ancient inhabitants of Belize, a people we call "the Maya," and the present inhabitants of Belize who are of Maya cultural heritage, the connection is weak" (Wilk 1986: 73).
29   On the wandering *milpa*, see Wright et al. (1959: 38, 132).
30   Ibid., p. 35.
31   Ibid., p. 125.
32   See Timothy Mitchell's outstanding study, *Colonizing Egypt*, where he shows how colonial disciplinary power "works not from the outside but from within, not at the level of an entire society but at the level of detail, and not by restricting individuals and their actions but by producing them" (1988: xi).
33   Ibid., p. 132.
34   The claim is groundless: cf. Wilk (1981). Southern Belize was in the 1950s, and remains today, one of the lowest-density regions of Central America. Nevertheless, neo-Malthusian arguments persist.
35   Wright et al. (1959: 1).
36   Ibid., p. 132. They go on to propose that the Maya farm system should be "refined in some way to ensure a more intensive use of the land without damage to the fertility of the soil." Elsewhere in the text Wright et al. acknowledge practices that maintain soil fertility in the *milpa*: "Shifting cultivation on hill slopes [by] Maya Indians [is] carried out on rugged limestone hills whose slope is broken by a considerable quantity of outcropping rock. Soil loosened by the rain seldom moves far down the slope. It is either caught in the pockets between the rocks or passes through fissures in the rock.... Under the present system of 'milpa' farming where the resting period between cultivation is from 4 to 7 years, *the limestone hill soils appear to be maintaining their fertility surprisingly well*" (Wright et al. 1959: 272, my italics).
37   Ibid., p. 134.
38   *Land* argues that the creation of reservations three decades earlier was a mistake because "it set the Toledo Maya apart from all the other racial groups... and much bad land use has resulted" (p. 271).
39   Ibid., p. 109. Full capitalization in original.
40   Ibid.

41 Consider the theory of development presented in *Land*, based on four "general principles" of development:

> (1) to "go carefully" so as not to work against "the forces of nature";
> (2) to make "a rapid inventory of the natural resources of the area involved";
> (3) to "ensure continuity of effort by securing the services of a capable coordinator"; and, above all,
> (4) to "keep people fully informed about plans for the development of their land" (pp. 107–8).

These points, however arbitrary, capture the hegemony of capitalism qua development today: sustainability, inclusiveness, trusteeship.

42 Ibid., p. 108.

43 Wright (1996a). No page numbers.

44 Ibid.

45 For Wright's cultural ecology, see Wright (1997a); see also Wright (ca. 1960, ca. 1994).

46 This is a reformulation of Bhabha's claim, cited in the epigram to the chapter.

47 Wright (1997b).

48 Wright et al. (1959: ii).

49 Heidegger (1977b).

50 Mitchell (1991).

51 Morley (1946).

52 Owen-Lewis (2001). All quotations by Owen-Lewis are taken from an interview conducted at his home in July 2001.

53 Nuñez notes that the perception that Owen-Lewis was "regarded as an eccentric" by other colonial officials, since he married a Q'eqchi' woman and became a "recluse." "Government felt that it saw too little of the Liaison Officer and in fact felt that it had neither knowledge nor control of his activities" (Nuñez 1977: 31; see also p. 44).

54 Ibid., p. 41. Nuñez asks: "Why is it that Government would spend so lavishly on an officer's living quarters yet allow the [boat] to rot for want of a few hundred dollars worth of repairs? Inconsistencies of this order occur in many large administrative systems, but they seem to abound in poor countries which can least afford them."

55 Owen-Lewis moved from Crique Sarco to Machaca, along the Punta Gorda-San Antonio road, in 1959. He was moved by colonial officials who felt that he was too inaccessible and living too modestly. As he explains:

> The Governor was staying in the village on an overnight.... I showed him what I was doing.... [He] got back to my little shack that I had in Crique Sarco, and he said ..., "you know, you should have some furniture here." "What do I want furniture for? I've got a hammock." "Yes, I know," he said, "but I've got nothing to sit on.... I'll get some money for you, and you can pick out what you want." I said, "Sir, before I get furniture, I'd really want a

better house to put it in." Because this was just a house [with] wood sides, thatch roof.... He comes in a couple of months later, and gets off the plane, and he's got a big roll of paper under his arm. And he said, "Don't forget, we've got to listen to the news at 6 o'clock tonight.... I am approving the site of your new house."... He takes his roll from under his arm, and stretches it out, and Jesus, this thing is a house, and an office block, and a Senior Officers rest house, and a clerk's quarters, and a laborer's lines, and this was a complex of buildings, costing a lot of money. So I said, "Where are you building this thing?" "Here in Crique Sarco." I said, "Jesus, we can't build this thing in Crique Sarco.... I've maybe only...another three more years, and then I quit, and then what happens?" He said, "Yes, you've got a point. Where would you want to put it?" I said, "Somewhere central in Toledo, so that when I bow out of the picture, somebody else can take it over."

Owen-Lewis only occupied the new building, the most elaborate state building in southern Belize, for a little more than a year. It became, and remains, the headquarters for the Forest Department in Toledo.

56   Nuñez (1977: 35). Nuñez writes that the "two most elaborate [structures], the Farm Demonstrator's quarters and the Dispensary with Nurse's quarters serve no purposes whatsoever. The most imposing structure in the community is the Police Station, which functionally plays only a marginal role in the village life."

57   This is not to say that Owen-Lewis placed great emphasis on his teaching the Maya how to be British *per se*. We should not confuse the practice of developing hegemony through British colonialism with the crude enforcement of Britishness. I asked Owen-Lewis if the Maya saw themselves as British subjects when he lived in Crique Sarco. He replied:

DOL   I don't think so. I don't think they saw themselves as Guatemalans when they were in Guatemala, either. They were just bush Indians.

JW   Is that something that you would try and...remind them of, or instill in them?

DOL   Oh, no, no, no.

JW   You didn't fly a British flag, [DOL: No.] or sing the British national anthem?

DOL   No, sir, I was too busy just trying to live – to live, and getting to know them.

There was no contradiction between Owen-Lewis's intense desire to *know* "the Maya" and his desire to transform them by linking them with British state policy and broader economic networks.

58   Owen-Lewis (1998).
59   Wright et al. (1959: 183–4).
60   Morley (1946: 158).
61   Ibid., p. 141.
62   "The distinction between the material world and its representation is not something we can take as a starting point. It is an opposition that is made

in social practice, and the forms of this opposition that we take for granted are both comparatively recent and relatively unstable" (Mitchell 2003: 6). See also Mitchell (1988).

63 Wilk (1997: 69). Owen-Lewis knew Wilk and Nuñez and felt their studies were accurate. He said of Nuñez: "he went through all the government papers and then he came down and he talked to me. And he was very fair. He sent me a copy of it afterwards. And I couldn't find too much fault with it at all; I thought it was very fair. [...H]e was a Jamaican, and he was a black man, and I thought if anybody's got a chip on his shoulder against a white, British colonial, ex-officer thing it would be him, but he wasn't. He was a very pleasant man." Owen-Lewis was proud to have received more commentary in Wilk's *Household ecology* than the TRDP project (see chapter 5), which cost millions of pounds but had, in Wilk's words, "little impact on the Kekchi economy" (1997: 71).

64 Nuñez (1977: 31).

65 Ibid., pp. 37–8, italics mine.

66 This is why the concept of Indigenous Technical Knowledge (ITK) is epistemologically colonial. ITK cannot be understood apart from liberal development critiques that argue that the problem with development is that it insufficiently grasped the local, indigenous knowledge as a resource for development.

67 I believe that Owen-Lewis recognizes this. Although he has remained in southern Belize, Owen-Lewis told me that he had not returned to Crique Sarco since 1960. He speaks of the Maya with nostalgia and a palpable anxiety:

> They were very law-abiding, exceedingly nice people. I can't say the same for them today. They're getting out of hand. They're getting spoiled. I see strife ahead down here. We've never had strife here before, but I see it ahead.... There is a lot of bitterness creeping in, and I don't know what's causing it. A lot of bitterness. It's part and parcel of this same syndrome that the more uncivilized they are, the nicer they are, and now they're looking at movies, and whatever. A lot of outsiders cause trouble too, they sort of build them up, and they make them think that they're something special and it kind of upsets the equilibrium. There is a lot of bitterness creeping in, and I don't know what's causing it.

As for the Maya movement, formed in part to prevent the privatization of common lands:

> There's going to be trouble. And it's going to be caused by the Indians. It's not going to be caused by anybody else. All of this trouble about a "Homeland" and stuff, you don't hear a peep out of the Caribs in Punta Gorda, you don't hear a peep out of the Coolies in Rancho. Nothing. They don't say anything, but they talk to me a lot. And they're really getting pissed off with this thing. And they're getting pissed off with foreigners coming in here and talking to the Indians and giving them advice.

Spoken by a colonial agent sent to give advice to the Maya, this is more than ironic. It is a tragic parable of epistemic violence wrought by colonialism.

68 Wright (1995: 5–6).

69 Ibid., p. 6. Wright adds that as a result of his recommendations, "much of the Crown Land of Belize was reclassified as Forest Reserve, and controlled timber and chicle production became a major source of income and employment." The last clause is a great exaggeration. Timber and chicle employment as a fraction of economic activity probably peaked in the mid-nineteenth century (see chapter 1).

70 Ibid.

71 When the UDP government faced a cooling economy during their second term in office (1993–8) – a period in which per capita economic growth was almost nil – they sought new sources of export revenue. Seafood (fish, shrimp, and lobster) and citrus exports were regarded as potential avenues for growth. Forestry came to be seen in a new light as well. Two concessions were granted for the CRFR, but were revoked after environmental NGOs exposed breaches of contract. Environmental NGOs argued that the CRFR should be logged very selectively, if at all. By 1990 several rapid ecological assessments conducted in the CRFR found it to contain habitat for a high number of endemic species. One study discovered 35 new plant species for Belize, only a few of which occur in other parts of Central America. The authors argued that the CRFR was of "special interest and not represented ... anywhere else in Belize and possibly not anywhere else in Central America.... Any change in land use could result in millions of years of tropical forest vanishing forever" (Meadows, M., cited by Wright 1996b: 8).

72 This subsection draws on Forest Department documents and interviews conducted with Forestry officials between 1996 and 2002 (see Wainwright 1998).

73 See also ODA (1989), which found that the Forest Department has "considerably fewer personnel, at all levels, in relation to the size of the forest estate under management and in use, than is normally accepted as being necessary" (p. vi). The key to profitable sustainable forestry lies in altering the given ecological conditions in forests in such a way as to increase the profitability of harvests over time. This typically entails selective logging and replanting of high-value or fast-growing species. After investment in the Forest Department fell in the 1960s, state capacity declined and forestry received also no investment or state coordination. The Forest Department is part of the Ministry of Natural Resources, but is governed by a relatively autonomous Chief Forest Officer responsible for managing the national forest system. The most important management decisions involve granting forest licenses. Decisions about licenses are supposed to be made by the CFO with the consultation of his staff. Much of the internal conflict

around the CRFR in 1995–6 and subsequent licenses in Toledo in 1996–8 resulted from the lack of transparency in the licensing process.

74 ODA (1989: iv). In a series of interviews conducted during July 1997, state officials explained that forestry was entirely divorced from national development planning. Details about pricing, volume, or exports are not linked to national economic plans; data on timber exports are not linked to data on revenue collection; the erratic shifts in forestry exports could not be explained.

75 McCalla (1995: 1).

76 In Toledo alone the Forest Department staff declined from six to three even as several new logging concessions emerged. Retrenchment left the department understaffed, stung by low morale, and consolidated under the power of the Minister of Natural Resources, Eduardo ("Dito") Juan, who enjoyed an unenviable reputation as a tyrannical boss. In 1995–6, Forestry staff expressed a palpable sense of powerlessness and concern for the condition of the department. Fear of losing one's job, coupled with the overwork produced through decades of underfunding, left the mid-level officials and forest guards with little confidence in their institution.

77 The TFAP led to the creation of no less than three new state institutions: a Conservation subsection of the FD; a rural development and conservation program funded by USAID (the Natural Resources Management Program, or NARMAP); and FPMP. Only the Conservation subsection remained beyond 1999.

78 Bird (1994).

79 The plan reserved 62 percent of the CRFR for conservation, but none for use by residents of nearby Maya communities.

80 Garcia (1999: 1).

81 Bird (1995). Unpublished memorandum to Richard Belisle, 2/28/1995, p. 2. Among other criticisms, Bird noted that the project proposal made no mention of "the social responsibility of the enterprise beyond providing direct employment" (p. 2).

82 Cho's talents were readily recognized, and he was subsequently elected Chairman of the TMCC on December 31, 1995 (see chapter 6).

83 The papers discussed the conflict as the "Malaysian" controversy. The "Malaysian" label originated because three of the on-site managers of the Atlantic Industries sawmill were Chinese Malaysians.

84 Cho (1995).

85 The Forest Department argued that "all logging operations within the Columbia River Forest Reserve has [sic] to be done in accordance with a comprehensive sustainable management plan developed with the assistance of forest management experts" (1995: 1; see also Forest Department 1996). The MNR claimed that there were sufficient resources to police the contract. The FD staff in Toledo knew otherwise; they had but three employees and one functioning truck.

86 One Forest Department press release claimed that "the greatest loss the forest has suffered in recent years is by agricultural expansion by local farmers" (1995: 2). At a press conference in 1996, Minister Juan alleged that *milpa* farmers were only interested in the CRFR as a site for marijuana production.

87 Bird (1995: 1).

88 Forest Department (1995: 2). Although the consent of local communities is required by the CRFR Management Plan, no alcalde had endorsed the plan.

89 Glaser (1995: 1). Glaser justified her conclusion on cultural grounds, noting that Mayas "do not express dissatisfaction initially but on further consideration and discussing become aware of implications with which they do not agree."

90 The FPMP officials were met at the entrance to the forest by an angry crowd of hundreds. The officials said that they had been sent to explain the details of the contract (road and bridge construction, boundaries, etc.) to AI, in the company of Maya leaders, in order to enforce the Management Plan.

91 When FPMP, humiliated by this revelation, failed to cancel the concessions, the Maya leaders decided go to the capital. In December, 1995, the RLC, TMCC, and TAA arranged for a busload of Maya leaders to travel to Belmopan. Their buses were stopped en route and they faced police harassment. The Prime Minister refused to meet their delegation.

92 TMCC and TAA (1995: 1), my italics.

93 This provoked a calculated response by the state. In January 1996, Prime Minister Esquivel called an emergency meeting with TMCC and TAA in Belmopan. Although their discussion resulted in no new agreements, the government announced that an agreement had been reached and the conflict was resolved: "The Government of Belize has agreed to a proposal by representatives of the Toledo Mayan Communities, with regards to logging operations . . . [which] . . . sets out certain guidelines and conditions [including]: the employment and training of Alcaldes in the surrounding villages to oversee the logging operation; to identify a specific block in the forest [*sic*] to be logged by the Maya . . . ; to provide the Mayas with expertise to help those who may choose to engage in logging; [to] assist the Alcalde Association to negotiate with Atlantic Industries Limited to make a quota of timber available . . . at cost" (Belize Information Service 1996: 1–2). With the bogus settlement threatening to divide the Maya movement, TMCC and TAA called an emergency meeting that resulted in the "Declaration of San Jose," in which they "categorically rejected [the . . .] concession granted to Atlantic Industries Ltd., [and] unanimously call[ed] on the government to desist from giving any logging concession in any areas of Toledo where the Mayan Peoples' harmonious coexistence with their environment is threatened" (TMCC and TAA 1996: 1).

94 Before long the government of Belize was criticized in US magazines and newspapers. *Newsweek* lamented: "In this age of international pressure to save the rainforest, promote sustainable development and honor indigenous cultures, the government of Belize seems to be ignoring all three" (Lucy and Koehl 1997: 4).

95 Cho and Coh met with the Inter-American Development Bank, the World Bank, the US State Department, the OAS Inter-American Commission on Human Rights, the Inter-American Foundation, the National Congress of American Indians, and various environmental organizations.

96 The case argued that the state infringed upon the property rights of the Toledo Mayas by granting logging concessions to non-Maya companies in areas of traditional Maya use and occupancy. The purpose of the lawsuit was to compel the state to recognize the aboriginal rights of the Toledo Mayas to the lands they use and occupy.

97 On the concept of social nature, see Castree and Braun (2001); Braun (2002); Castree (2005).

98 Data on office visits comes from Wright's careful visitor records of 1995–8. Wright's approach to the conflict influenced Cho, and no doubt Cho influenced Wright as well. In an essay from this period, Wright writes about Maya land requirements and population:

> In 1952, the Maya Indian population was estimated to amount to 5,000 families (which we soon found to be a considerable under-estimate). Each family was said to require 4 to 6 acres for annual corn (maize) production, plus another 50 to 60 acres which would gradually be used for milpa crops at a rate of about 6 acres each year, allowing each milpa abandoned after one year's use, and adequate time to recuperate soil fertility under regenerating forest species which was thus, in effect, a "bush-fallow" interval.... A population of 5,000 families thus would require that about 500,000 acres of the landscape of Toledo would be needed for production of subsistence foods. (Wright 1997a: 22)

There is a strong parallel between the figures Wright provides here and those offered by TMCC's proposal for land reform. In 1997, when Cho and the TMCC proposed the creation of the Maya Land Administration Program, Cho argued that the amount of land that would be necessary to provide all Toledo Mayas with sufficient land for food production was 500,000 acres. In each case, the number 500,000 represented a claim: with Wright, it is the spatial unit corresponding to the "ecological footprint" at the point where Maya livelihoods match the land's carrying capacity. Cho transforms this number into a claim to territorial rights.

In the same text, Wright notes that the land use practices of the Mayas could not be mapped in a way that represented them as fixed, demarcated, or permanent:

> When we started to walk across the landscape ... we soon learned that a Mayan family "farm" had no fixed boundary. A farmer could select land for

> his annual milpa in one location, [and] move on to a new location often at some distance from his old location for his next annual milpa.... We only very rarely found that a Maya family had kept their farming activity close enough to the starting point to constitute something approaching the European concept of a "family farm." Land was regarded by the Mayas as a communal resource and each family had the traditional right to move.... We thus quickly learned that no part of this flexible system of land utilization could be used to make a credible map of land use. (p. 22)

In his growing awareness of the limits of his earlier knowledge we can detect a desire to decolonize the form of development knowledge that he helped to produce.

99   See especially "Charles Wright, OBE, on the raping of Maya ancestral lands," *Amandala*, August 17, 1997. Wright, "The Lamanai Room Declaration: A reply from Toledo," September 8, 1997. AB, BAD/CHW/28; Wright, 1997 ca., "Approximately the current activity in the areas with forest licenses in Toledo." AB, BAD/CHW/26. He also prepared several drafts of a text on the history of land use by the Maya in southern Belize covering two thousand years (Wright 1997a).

100   Wright (1995: 1).

101   Ibid, p. 2.

102   Wright (1997a: 33).

103   Wright (1995: 3–4).

104   Ibid., pp. 3–4.

105   Wright's essentialism was one of the main reasons that the CRFR conflict was framed as a "Malaysian" problem. He always identified AI as a "Malay" or "Malaysian" phenomenon, even after learning that the company was based in Hong Kong.

106   Ibid.

107   For instance, Neil Bird, the head of the British forest project that advised the Forest Department during the conflict over the CRFR, wrote a book about his work in Belize in which he asserts that his management plan for the CRFR "has been put into practice and found to work well" (Bird 1998: 125). The intense and conflicted sociopolitical life of the forest is simply repressed. The text bears no mention of the conflicts that raged around the forest; the noise outside the lab never enters his report.

108   After all, most others – including Owen-Lewis – never moved as far as Wright. When asked about the conflict in the CRFR, Owen-Lewis said:

> If I'd been a bloody Indian, I'd a done what Columbia people did way back when must have been 1960. [A logger] had that little woodworking thing up on San Antonio road. He came in with a couple of bulldozers and started working by San Pedro Columbia. San Pedro Columbia had just fixed themselves a farmer's road, and put up some little wooden bridges so that mules could cross and people could cross. Along comes [the logger] with big tractors and stuff and starts crashing these things and breaks them up and

screws up the road, and the Indians said "Hey, you're screwing up our road. You're making ruts in our road and we just fixed it." [He] said, "Well, you fixed it with government money, so I've got as much right as you have to use it." So this went on for a little bit, and in the end the Indians, well, they burnt one of his bulldozers and two of his skidders. They set fire to them – outright burned them.

This is tough talk, but the intensity of it should not throw us off from recognizing that Owen-Lewis, who never publicly opposed the logging in the CRFR and celebrated the death of Julian Cho, says what *he would have done*. This is a mode of speaking for.

109 This is not changed by the fact that Wright's own texts were a part of the colonial archive – since those texts also produced Wright. This is not to say that he was trapped by his own voice, but rather that his texts both reflect and contain the limits of a particular way of being and speaking about the Maya.

110 There rarely seem to be *only two* valences pulling at the colonial texts. Texts bifurcate, trifurcate, unsettle, and combine to the strings of multiple desires; an overemphasis on ambivalence can make our readings of the colonial archive too neat. What is at work in colonial discourse is not only the (mis)identification with the other, but rather the becoming of anticolonial positions within spaces marked by colonialism, capitalism, nationalism, development hegemony, sexism, etc. These surfaces (I do not say oppositions) may scatter the subject or text as multiplicities in ways that turn on more than one dynamic. Ambivalence is therefore perhaps too limited a means to understand the valences in the colonial archive.

111 Wright, ca. 1994. "The digging stick (thoughts about enhanced integration of the Toledo District with the rest of Belize that will follow from up-grading of the Southern Highway)." Unpublished manuscript. AB, BAD/CHW/31.

112 Wright was becoming aware of this in his later years, and towards the end of his life he began working on a history of all the development projects that had come to southern Belize since the 1950s. His draft version of this essay is one his most important unpublished texts; see Wright (1996b).

113 On CARD, see the introduction to chapter 2.

114 Said (2002 [1980]: 97).

5

Settling: Fieldwork in the Ruins of Development

colonize. *v.* **1.** *trans.* To settle (a country) with colonists; to plant or establish a colony in.... **3.** *intr.* To form or establish a colony or settlement; to settle.

Oxford English Dictionary

On a torrid Sunday morning in November, 1995, I clambered into the back of a rented pickup truck driven by Julian Cho, and we drove unpaved roads through five villages to collect the members of the Reservation Lands Committee. From our last stop, a house near the old church in San Antonio, we turned back to Mafredi to circumvent the *witzil cortez* and crossed into the floodplain of the Moho River at mid-reach, where rice-fields with high grass weeds appeared in breaks of the secondary forests. The Reservation Lands Committee (hereafter, the Committee) soon arrived in Blue Creek for a village meeting, to conduct research on land tenure and mobilize the community. Propelled by a mandate from the alcaldes, the Committee had organized a campaign to discern the status of land tenure in every Maya community in the District and mobilize the Maya on the basis of the accumulated knowledge. The patchwork of land tenure regimes inherited from colonialism was rapidly fragmenting, and knowledge of the local land situation had to be gathered directly from each village. The Committee went to mobilize the Maya communities while conducting subaltern geographical research: research and politics merged as fieldwork.

With funds granted by USAID to advance forest conservation,[1] Cho and his Committee would visit two communities each Sunday after church and present the message that USAID wished to spread – forest conservation requires secure private property rights – and then unsettle its presuppositions. Cho and the Committee argued that the Maya communities must build alternatives to both the colonial Indian reservations and also to privatization, by way of a struggle for collective, indigenous land rights.[2] The conflict over the Columbia River forest (see chapter 4) illustrated the dangers of the present situation. Without indigenous land rights, how could poor communities prevent the theft of their resources by outsiders?

This was my third Sunday excursion with the Committee, but my first time crossing the green waters of the Blue Creek, climbing its limestone river bank and passing the football field, tracking down the alcalde, and arranging the meeting. When the plans were settled, the Committee divided into two groups and sent one to a neighboring village, Aguacate. As we drove westward from Blue Creek, we passed a curious assemblage of decaying structures, an unexpected complex of dilapidated buildings that lay open to the sky, like a damaged and abandoned city. Around the buildings were strewn agricultural machines – broken, hulking piles of rust and thick with weeds. One of the Committee members explained: "That was IFAD, a big development project. And now" – gesturing to the ruins – "someone could do archaeology here."

The fieldwork that culminates in this chapter started that afternoon, when I learned with the Committee from the farmers in these two communities that their lands, part of an Indian reservation until the early 1990s, had recently been subdivided by a development project. When their lands were parceled out, the farmers did not receive papers for land title: the land papers went to the Development Finance Corporation (DFC), Belize's national development bank, which held the land as collateral for loans. The farmers explained that they did not receive cash from the DFC, but rather agricultural inputs: orange saplings, cattle, and assistance in land preparation for mechanized rice production. Promised agricultural and marketing assistance never materialized, however, and the farmers were left with oranges they could not transport, mechanized rice they could not harvest, and cattle they could not sell. Already by October 1995, the farmers in Blue Creek and Aguacate had fallen behind in their debts and had ceased paying the DFC for their loans. They told us that the DFC was threatening to foreclose on their lands, and pieces of their reservation would be sold to outsiders. The farmers asked Julian Cho and the Committee to take their plight to the government, to fight to get their debts cancelled and land papers back.[3] In July, 1997, Julian asked me to return to these communities and report to him on the land situation and options for mobilizing the communities. This chapter comprises my report to the late Julian.

Yet my report is neither an exposé of the bad management of effects of development projects, nor another iteration of development discourse analysis. Rather, this chapter reports on fieldwork in the ruins of development. Though some transcoding is inevitable, my "fieldwork" in these communities has not been oriented toward data collection and knowledge production.[4] Fieldwork is usually executed to solder a circuit that carries empirics from *there* (the field) to *here* (the center of calculation). Here my model is closer to what Spivak has called "fieldwork without transcoding," a kind of thinking that uses a "patient effort to learn without the goal of transmitting that learning."[5] This does not imply a refusal to abide by or study in what we conventionally call "the field" – on the contrary. Neither is it a facile rejection of collecting subaltern geographies or histories. Rather, it is a way of performing fieldwork to unlearn, to learn to read anew, to learn apart from knowledge-production in the mode of empirical data collection and transcoding. These fieldnotes reflect, I hope, something of that work. They aim at something I learned "in the field": the importance of unsettling the metaphysics of settling that structure capitalism qua development.[6]

Settling In

In 1978 a development project came to two villages in southern Belize. In the course of this project, called the Toledo Resource Development Project, or TRDP, and a subsequent project called the Toledo Small Farmers Development Project, or TSFDP, roughly five million dollars were spent in poor, rural Maya communities to promote – in the words of the sign advertising the project outside its headquarters – "settled agriculture."[7] The core project documents identify the goal as "the development of more settled systems of farming."[8] Success in these terms was defined by shifting Maya farmers in the region from *milpa* maize agriculture to "settled crop systems," especially mechanized rice production. This would decrease forest clearing, increase rice production, and improve Maya livelihoods.

Measured in terms of their stated goals, these projects were abject failures: Maya rice production did not increase; Maya farming practices did not markedly change. Nor did these projects contribute to capital accumulation in the region, which remains among the poorest in Central America. And within two years of the end of the second project, the

elaborate buildings that were constructed to house the expatriate development experts lay in ruins. Yet these projects have had concrete, unexpected effects on the communities they visited: the parceling of formerly
commonly managed community lands; the indebtedness of many farmers;
the rise of the Mennonite church in a historically Catholic region; and
more. Rather than "settle" Maya agriculture, therefore, we may conclude
that these development projects *unsettled* Maya livelihoods.

As the OED reminds us (see the epigraph to this chapter), the essence of
colonialism lies in settling: the spatial practice of moving to a "peripheral"
space and making it home. Just as colonialism entails much more than
simply settling in a foreign land, the verb "to settle" aligns a vast and
heterogeneous set of meanings. The OED discerns 36 meanings of the
verb "to settle," including: to seat, place; to place in order; to take up
residence; to come to rest; to establish a fixed abode; to descend, lower; to
bring to rest after agitation; to become composed; to come to a quiet or
orderly state after excitement or restless activity; to subside into indolence
or contentment; to compel to cease from opposition or annoyance; to
silence; to fix or become fixed in a certain condition; to ensure the stability
of; to secure or confirm (a person) in a position of authority; to secure
(payment, property, title) to, on, or upon; to subject to permanent regulations, to set permanently in order; to appoint or fix definitely beforehand; to come to a fixed conclusion on; and to bring to an end (a dispute)
by agreement or intervention. In short, to settle is to resolve difference,
bring to accord, end agitation, in such a way that achieves spatial fixity
and stability. To settle is to sediment sociospatial relations. Thus "to
settle" is a synonym of "to win colonial hegemony." As we have seen in
Part I, the key verb in the colonial archive concerning the Mayas was "to
settle." The "Indian problem" in Belize, restated over and over, has always
been that the Maya are not settled; they are scattered, unfixed, indolent, in
dispute, un-placed, agitated. The problem is literally tautological: define
a people as unsettled, and the solution to their problems is settlement. Fix
them. Pin down the wandering *milpa*. This fixed conclusion is constituted
through one particular settlement: capitalism qua development.

Settling the Maya: The Toledo Research and Development Project

Development projects, too, weave origin stories, always multiple and
often contested. Typically their histories narrate a setting – a space-
economy – and the story is that a region needs development because

the economy is badly placed or mismanaged.[9] By defining a space-economy as an object in need of development, development texts provide setting, i.e., they set the scene for capitalism qua development. Open the original project document of the TRDP from 1977, and you will find that the aim of these development projects from the outset was the transformation of the Maya farm system via settlement. The setting for this particular narrative is the margins of the colony on the eve of formal political independence: TRDP started in 1978, and independence came in 1981. For the nascent Government of Belize, TRDP gestured to an interest in "develop[ing] the Toledo area more fully, not only to enable more food to be produced and to raise the standard of living of the people, but also to encourage...the Maya Indians...to identify themselves more permanently with the rest of the country."[10] The unfinished colonial project of territorializing southern Belize as such – reflected in the failure of "permanent" national-territorial identification – gave rise to the need for settled development. The post-independence state desired territorialization, the binding of nation and territory: "The Government therefore wishes to encourage the development of more settled systems of farming in the region. Thus the idea of establishing a special project [the TRDP] to investigate current practices and evolve new patterns of settled farming and rural development was conceived."[11] This union would be forged through capitalism qua development, born of a political desire for national identification: "government policy has stressed the development of agriculture, especially rice cultivation, as a means of *drawing* the Toledo District closer to the aims of national development."[12] The Maya would be settled to identify with property and the nation.

Nationalism always implies erasing certain differences, and development was intended to articulate the Maya within national territory – not reservations. The TRDP's earliest documents explain that the project needed to advance the new Belizean state's intention "to do away with the [colonial Indian] Reservations" – notwithstanding "considerable local opposition" to dereservation.[13] Development meant a directed change – greater incorporation into markets and privatization of land – with uneven consequences:

> Land and Maya welfare are inextricably interwoven, with much of their social activity being organized around the tempo of the agricultural cycle. Most Maya have access to land on reservations, but legally they do not own it, having only usufruct. Government policy is the gradual

eradication of reservation land in that it is felt to discriminate against non-Indian users. There is a fear that dereservation would deprive the Maya of much of the land near their villages having greatest agricultural potential.... Maya farmers are said to be unfamiliar with lease arrangements. Safeguards would need to be devised for those who are constrained by tradition, otherwise land reform might have very divisive and damaging consequences for the Maya. The more adaptable may become prosperous independent farmers. Many more might become impoverished landless laborers eking a livelihood from declining "milpa."[14]

From the outset, then, TRDP documents note the risk of driving subsistence farmers off the land – of proletarianization. Project planners knew that the Maya farmers were capital poor. The solution, then as now, was to facilitate the conversion of common property into private holdings: to free up the land that the community managed collectively as capital, so that farmers could take out loans needed to get into mechanized rice production. Here is this "capital logic" as described in 1981: "none of the farmers would have the capital to buy the land initially.... The farmer would, however, be given title to the land from the start; this would enable him to use the land as a security for a machinery or development loan."[15]

The problem with this strategy is plain. Why would a poor community convert its secure, commonly managed resources into private property which may be divided, sold, lost? To put the same question differently: why change the livelihood strategy that had settled the community? Here is how the development experts pose this question:

> The Maya ... are essentially peasant farmers working under a "milpa" or shifting cultivation system of agriculture.... The "milpa" system has sustained the present standard of living, required few inputs, and involves little risk.... Moreover, cultural constraints may militate against innovation; some evidence suggests that Mayan values deprecate material acquisition and ostentation. While the "milpa" system will sustain the present way of life, it has limitations ... [including] land tenure due to its transitory nature.... [T]his policy will have profound significance for [Maya communities]. If the Maya are to benefit they must assimilate much new knowledge about both agriculture and government administration. Similarly, government['s] stated intention to regulate land tenure, and to opt for larger settled farm units ... will demand significant cultural adaptation on the part of the Maya.[16]

Unfortunately, the experts note, once population densities rise to a certain point the fallow period decreases, productivity declines, soil degrades. Now, at the time this was written, the population density of the Toledo District was the lowest in Belize, which had in turn the lowest population density of any country in Central America.[17] As for land use: according to data from the 1973–74 agricultural census and TRDP's land use survey of 1983–84, the mean number of acres farmed per Maya family in Toledo declined from 6.8 in 1973–74 to 6.4 in 1983–84.[18] The TRDP documented no substantial declines in fallow or productivity, either. But no matter. To redirect the fate of the Maya farm system requires an intervention to facilitate the development of the Maya farm system. The intervention will be made from outside of the Maya farm system by a trustee: the colonial state. In this sense, the Maya farm system solicits its own trustee.

On the eve of political independence, this reformation of development trusteeship took on formal political qualities in electoral contests. Although George Price and his People's United Party (PUP) had captured power in the early 1960s, all through the 1970s they fared poorly in Toledo.[19] In the elections of 1974 and 1979, Toledo was the only political district to elect two opposition United Democratic Party (UDP) representatives – and in each election, one representative was, for the first time, Maya. The PUP's sense that it had failed to win nationalist hegemony in the south therefore mapped onto the region's poverty, ruralness, and indigeneity. Moreover, this region was (as it remains to the present) claimed by the Guatemalan state as part of its territory. Particularly in the years before 1978, before the onset of the civil war in Guatemala, many Mayas in Toledo questioned their potential gains from an independent Belizean, PUP-led government, even going as far as to endorse the Guatemalan claim. Price and the PUP saw a need for a major development project that would signal its commitment to developing the rural south while articulating the Maya into their national development program. One report suggests that Price was pushing for the TRDP as early as 1975, since "there is still strong anti-Government [i.e., anti-PUP] feeling" in Toledo, and "there is a national political desire to do something...for the development of Toledo...to spike the guns of those who advocate better economic progress through union with Guatemala."[20] Development was therefore a means to tie the nation's fragments to its territory.[21]

Development's abstract relations of power must touch down, of course, and their resulting effects are influenced by the varied and subtle geographies of soils and streams, religion and social class. Consequently the geographical work of settling development projects in particular spaces is always significant. In 1976 a survey was conducted to examine "the social factors that would be involved in settling families on Mafredi swamp," focusing on "the Indian population in settled agriculture."[22] The author, Richard Holloway, based himself in San Antonio and visited ten rural Maya communities. But his research did not begin in those communities. To prepare for this assignment he read a forty-year old ethnographic study of the Maya of Belize – J. E. S. Thompson's *Ethnology of the Mayas of British Honduras* (1931) – and *the* obligatory passage-point of colonial development, Wright et al.'s *Land in British Honduras* (see chapter 4). These texts frame and inform his important text on the "social factors" of the region and subsequent planning.[23] Holloway's report reiterates Wright's view of the Mayas as participants in an unsettled farm *system* that calls out for trusteeship: "The Kekchi are semi-nomadic people who not only farm the milpa system but frequently move their villages to new locations."[24] What needs to be done to Maya agriculture "is already outlined" in *Land in British Honduras*, he explains, "but has never yet been done," viz.: to conduct research that will determine "a supplement to and then a replacement for milpa cultivation."[25] The wandering *milpa* must be settled.

Thus when the TRDP was established in 1978, it materialized the inherited colonial discourse on the Maya farm system as an object in need of development. Its stated aim was to increase agricultural production by promoting a change from shifting maize production to settled farming. The underlying goal was to articulate the rural Maya communities into the national-territorial economy. Development was comprised of privatizing landownership, increasing the capital intensity of agriculture, and deepening a national political identification. Yet most Maya farmers had no capital to invest, so the project aimed at creating a major rice-growing center and securing individual land title.[26] The location for the research project – an area between two long-settled Maya communities, Blue Creek and Aguacate – was selected for its hydrology and the presence of the soils suitable "for the establishment of a mechanized rice paddy scheme."[27] It would not be incorrect to say that the project was imposed on indigenous communities with longstanding ties to this land and livelihoods based on common land management.[28] The TRDP intersected with livelihoods that were not simply indigenous and pre-colonial, but also the products of centuries of colonialism, displacement, and capitalist social relations. Although the communities and the TRDP were within colonial Indian reservations (see chapter 1), local practices had, before the TRDP, managed the village lands of Aguacate and Blue Creek as a commons.

"To Achieve a Settled Population"

That things went badly for TRDP is undisputed. The project documents give evidence of trenchant internal criticisms that alone raise fundamental questions. Consider, first, the ruins that the Committee saw on our trip to Blue Creek in 1995.[29] Only ten years earlier, these same buildings were at the center of a major controversy – and arguably the TRDP's most concrete achievement. TRDP's British funding agency, the Overseas Development Administration (ODA), felt it important to build elaborate on-site housing, replete with modern conveniences, for its expatriate staff.[30] But it took three years and enormous sums to build the houses, delaying the entire project. For the first two years on the project, the agricultural engineer charged with developing the new mechanical rice system spent most of his time constructing houses.[31] The housing construction consumed the project, and the work-lives of nearly all involved, and created immense tensions about delay that lowered morale.[32] Only six houses were completed, and they formed what could only be described as a segregated community: a complex of comfortable concrete houses with running water and electricity next to, but outside of, a Maya community without water, electricity, or the comfort of a waterproof roof. No housing of any sort was built for the subaltern, Belizean staff – who mainly lived in Punta Gorda – thereby creating "a real problem [for] local counterpart and junior staff" who were forced to commute to a remote community with no regular bus traffic.[33]

The British development experts worried over the construction until it was finished. Then they grew anxious that the newly independent Government of Belize would not appreciate their inheritance. This reflects TRDP's second major internal criticism: that the British and Belizean states were at odds over its purpose and direction. Nearly all of the texts in the TRDP archive were written by the British experts who designed and managed. Their texts reiterate older tropes (fixing the Maya, saving the forest) that were undoubtedly keenly felt, though we could speculate about economic motives for their investments: A. C. S. Wright, for instance, once claimed that the "TRDP was created mainly to provide 'jobs for the boys,' otherwise unemployed at that time on the over-large staff of the Overseas Development Ministry."[34] No doubt the departing British were anxious for a smooth handover and good relations with the new state. The British attempted to secure the government's commitment to the project. In April, 1981, four months before Independence, the government replied to one of the BDD's inquiries through the Ministry of Natural Resources. The Minister clarified the government's view of the purpose of the project:

> The policy of the Government is directed at stabilizing the land use situation in the Toledo District thereby making it more practicable to bring the benefits of agricultural technology and social services to the people.... Government considers that the best method to stabilize the situation is to provided [*sic*] freehold land titles to all those persons now farming in the area.... Government wishes as a first measure then to achieve a settled population in the area with adequate land under freehold title to secure a reasonable standard of living.[35]

As an afterthought, the Minister of Natural Resources adds that the TRDP project is "responsible for answering the likely social and related problems which might arise in the transition to a more settled way of life."[36] The new government, that is, asks the departing British to assume responsibility for the consequences of settling the Maya.

The tension between the two states over the effects of capitalism qua development fractured the anxious TRDP staff. Three years after TRDP's initiation, the project director wrote that "there has not yet been any real dialogue about development and its implications" between the project and the state.[37] The Government of Belize promised to appoint "a counterpart project manager" at the end of "the building phase,"[38] but none was ever appointed. Given the morass of elaborate house construction and half-finished agronomic research projects, the state was justified in hesitating to commit. Yet without state leadership, the TRDP floundered and the state's involvement declined further. The ODA therefore considered cutting off funding as "the project has been kept seemingly isolated from decisions taken by the Belize Government," but continued to fund the TRDP through 1986.[39] Research priorities were shifted "towards the uplands," away from Blue Creek and mechanized rice, because many "questions remain to be answered particularly concerning rice growing, and it would be folly to terminate prematurely a segment of the programme which is at an advanced stage."[40] Nevertheless the evaluation warned that land tenure remained "a fundamental issue," and even if it could be addressed, "a switch to growing of rice as a settled form of lowland farming... is unlikely to appeal to the Mayans unless the financial inducement is high and/or pressure are such as to promote a change."[41]

While these states clashed, one might ask: what of the views of the Maya? What of their voices about the object of development – their farm system? Of course they did not have an opportunity to criticize the project. When Spivak wrote "the subaltern cannot speak," she offered a potent tautology, and its productivity can be seen here.[42] The Maya were subalterns to Price, the ODA, and the new Belizean Government. They spoke, they resisted, but neither the state nor the TRDP would hear them.

Development projects are not often open or democratic affairs, and the TRDP was no exception. The original project report, written in Barbados by British experts, gives no indication that any Maya people were consulted about the project. The subsequent discussion about the plan was limited to London and Belmopan and seems to have involved no Belizeans save for the Premier, George Price, and the Minister of Natural Resources. Although it centered from the outset on the trans-formation of the Maya and their farm system, no Maya people were employed or held in positions of any authority. When thirty copies of the 1982 project mid-term report were distributed, not one went to a Maya person. Of the 72 people present at the final workshop, only two had Maya surnames.[43] The project was not of, by, or for the Maya: the TRDP was a project performed *on* the Maya, directed *at* them.[44] And when the unfortunate consequences of the project arrived, the Maya would be blamed for them by the state.

"An Honest Report of their Failures"

As the TRDP came to an end, far short of its goals, the expatriate staff knew they had failed. Their internal texts adopt a kind of development project "late style": more speculative presentation of ideas, greater use of vernacular, and a contemptible optimism that redefines project success and finds hope in any vague step in "the journey towards modern-ity." The recurring theme in the archive of this period was "there is a long way to go to find a system that could be adapted by the Maya farmers."[45]

Internal assessments of development projects may be critical, but they invariably nod to the brightest qualities. Two cultural anthropologists note (after summarizing the problems with the project) that the TRDP "yielded some beneficial results when those in charge realized that their attempts to drastically overhaul traditional Mayan agriculture were doomed to failure."[46] Yet it is telling that they give little details about these "beneficial results." Within Blue Creek and Aguacate, the TRDP was generally remembered with some fondness – not because of its

lasting consequences but as one of the only wage-paying employers ever
to visit these communities. A 1983 report by TRDP's agricultural econo-
mist notes that one potential source for income that could be used to
purchase inputs for rice farmers is "casual employment at TRDP."
"Some 500 'three week jobs' are taken up each year which earn the
recipients an average of $250 per stint."[47] Here was a development
project that spent so much capital, in such a small area, that it could
almost produce an accumulation project in itself, hiring enough farmers
that they could afford to make a transition to a more capital-intensive
production process – also provided by the project.

A. C. S. Wright offered the sort of pithy, frank assessment we may
expect from an ex-colonialist: "at least TRDP left behind an honest
report of their failures." The records of the TRDP were indeed truly
massive – it was the first development project in Belize that employed
computers for the everyday work of data crunching and memo writing –
but they, alas, were biodegradable.[48] But one cannot claim that the
records themselves could speak: there are no complete accounts of the
project, and arguably the most "honest report" of the TRDP's failure is
that the project now lies in ruins. And Wright, of course, is implicated in
the TRDP's failures. We saw earlier that his 1959 text framed the
original project document; it was also one of only four texts cited in
an important 1978 study of the likely social implications of the TRDP
(where Wright is named as one of three "particularly fruitful" local
experts who advised the TRDP's planners).[49] It is thus revealing to
read his thoughts on the project, written in 1996:

Some twenty years after *Land in British Honduras* was published, it was
noticed that the forgotten district of Toledo was not contributing much to
the national economy. . . . It was suspected that continuing growth of the
population within the various Mayan Indian Reservations was putting
pressure on available land resources and the appropriate solution would
be to encourage a more settled system of farming which might lead to a
greater production of cash crops [as Wright had advocated – JDW].
Accordingly, in 1978 an approach was made to the British [ODA] for
technical assistance. The target for this assistance was clearly the Mayan
farmer and, as a base of operations, a "pilot farm" site was located near
the relatively new village of Blue Creek in a lowland situation, but just
outside the hilly terrain where most Mayan milperos live and grow their
crops. . . . A lot of time was spent by the agronomists in trying out new
varieties of crop plants on the research station, but few survived field
testing on the farmer's own land. New weedicide, pesticides, etc., and a
variety of fertilizer combinations were successful on the research plots; but

on the farmer's land, any improvement in crop yields turned out to be insignificant in comparison with the costs involved. Much work was done on mechanized rice production on the wet soils of the lowland adjacent to the station but this was abandoned when adequate control of water proved very difficult. Even if this had been a successful venture, the [TRDP anthropologist] warned that few right-thinking Maya families would have considered leaving their life in a hill village to take up solitary existence in a swamp.... [T]he Mayan farmers gained very little from eight years of TRDP operations.[50]

This succinct abstract of TRDP's failure begins (correctly, as we have seen) with reference to *Land in British Honduras* and ends with the swampy matter of the desires of Maya people to be settled by others. Notably, Wright's narrative repeats the claim that the TRDP fell short of its goals because it did not understand Maya livelihoods well enough, because it did not conduct enough anthropological research on the Maya farm system. Wright reiterates his precept: do not try to develop the Maya before you understand them.

The Promise of Development Ethnography

In the Introduction we saw that the compound sign "development" refers simultaneously to an ontological process (unfolding of essence) and an externally directed practice (willful change). The affiliation of capitalism with this twofold sign has not simplified the responses to the practical problem of designing a metric for development. There can be no correct measurement of development without first settling a prior, indeed fundamental, question: "What constitutes development?"[51] Time and again, development theorists and experts have turned to one particular form of colonial knowledge to assist with these questions: cultural anthropology. Since its inception, cultural anthropology has professed a means to discern reason and values in the cultural practices of other, non-Western peoples.[52] Often this was to contribute to colonial hegemony, which most anthropologists today would criticize. And yet anthropologists continue to be solicited to serve as experts in discerning cultural change (prospects, barriers, and effects) for development projects that have taken up the colonial mantle. If the essence of colonialism is settlement, then the settlement of capitalism qua development has always relied upon some means of measuring cultural progress in the other – a task which calls for cultural anthropology.

In 1979, with TRDP held up for lack of housing and literally stuck in mud during long commutes, the project hired a cultural anthropologist: Ann Osborn, a British anthropologist then living in rural Colombia.[53] Her role was to provide the project with "a means of monitoring the progress made against the backdrop of Maya culture" in order to "assist the Maya in the assimilation of new ideas and technologies." Her first step was "to construct a broad ethnographic picture of Mayan life," beginning with a "detailed understanding of the organization of the 'milpa' system" and "inquiries into customary land tenure practices as a means of assessing Mayan susceptibility to change to more settled farming patterns."[54] Osborn was also charged with handling community relations, including labor disputes and village boundaries.[55]

The earliest records of her work with the TRDP are letters she wrote fellow ethnographers and Mayanists. In those letters, she defines herself as "a social anthropologist... involved in 'development,'" employed to document "the social organization of communities that may be affected by the project and in particular to record the present agricultural system of the amerindians and point to the social consequences of change in farming practices."[56] In another letter she explains that she is "documenting milpa, as it is today." Osborn thus defines the two tense elements of her subject-position: cultural anthropologist on one hand, development expert on the other. Osborn identifies strongly with the former and expressed no qualms about representing the Mayas as an anthropologist. Yet she maintains an identity as a skeptic of the potentially harmful consequences of development for the Mayas. Thus – much like post-development scholars in the 1990s – she nearly always set the word "development" within scare-quotes. For instance: "I am a social anthropologist and... have become involved in 'development.'"[57] Note the variation in this sentence: Osborn *is* an anthropologist who *has become involved* in development. She identifies with her discipline, not her employer. Her ambivalence is not about ethnography, but "development": "I am now working as a Social Ant[h]ropologist on a 'Development Project.' My work is very much involved in the problem of milpa cultivation."[58] "Development" is scare-quoted, but not "the problem of milpa cultivation."[59] Development is abstractly problematized, yet concretely employed.

Like the experts who designed the TRDP, Osborn initiated her research by reading Maya ethnographies. She built contacts in different communities, soliciting native informants (whom she paid) and studying Maya cultural practices. On this basis, Osborn wrote a collection of ethnographic reports on nutrition, social organization, health care, and education; each aimed at defining what is essentially Maya and indicating areas for development. Applying the tool of ethnography to the project's target population, she aimed to temper the negative consequences of development. If Maya culture was a barrier to change for the development experts, for Osborn it held the key to understanding its limits, flexibility, and management. Just as Wright's precept teaches, *knowing* indigenous practices is the basis for *predicting* social consequences. This reflects Mayanism (the Maya are what they are as ethnos), but also Eurocentrism: Osborn, after all, turned to the European ethnographers who could explain the Maya.

By questioning whether the privatization of the reservations and mechanization of rice production would be in the best interests of her Maya informants, Osborn caused problems for TRDP. Her arguments in meetings and memos, dotted with results of her ethnographic studies, set her at odds with the project's leaders: where the agronomists saw a farm system in need of replacement, she saw a functioning, if somewhat fragile, cultural system. She argued that the Maya farm system was less destructive and more efficient than assumed by the TRDP agronomists, who lacked data to argue against her. This ruffled the feathers of the other TRDP experts – male agronomists and economists – and she sensed that she would be forced out. In 1981 she wrote to a friend: "I do not know the direction of my part of the project – being realistic we all know that it is always the 'problem' disciplines (those that are willing to take up fundamental issues) [that] go first!"[60] When her contract ended in early 1982 it was not renewed.

Osborn's experience with the TRDP raises many pertinent questions, but I will focus on the one raised by her claim in this letter: exactly how big a problem is anthropology for capitalism qua development? Is anthropology a "problem discipline" at all – one, in her words, that is "willing to take up fundamental issues"?

Twenty years ago, Partha Chatterjee provided a lapidary analysis of the relation of power intrinsic to anthropology:

> No one has raised the possibility, and the accompanying problems, of a "rational" understanding of "us" by a member of the "other" culture – of, let us say, a Kalabari anthropology of the white man. It could be argued, of course, that when we consider the problem of relativism, we consider the relations between cultures in the abstract and it does not matter if the subject-object relation between Western and non-Western cultures is reversed: the relations would be isomorphic. But it would not: that is

precisely why we do not, and probably never will, have a Kalabari anthropology of the white man. And that is why even a Kalabari anthropology of the Kalabari will adopt the same representational form, if not the same substantive conclusions, as the white man's anthropology of the Kalabari. For there is a relation of power involved in the very conception of the autonomy of cultures.[61]

Chatterjee's critique is a postcolonial one, but we could also say that it is post-structuralist, one that extends Derrida's 1966 critique of Lévi-Strauss's structural anthropology.[62] In that essay, Derrida writes that "ethnology could have been born as a science only at the moment when a decentering had come about: at the moment when European culture... had been *dislocated*, driven from its locus, and forced to stop considering itself as the culture of reference." Like Mayanism, anthropology emerges in the mid-nineteenth century at a "moment when European culture... had been *dislocated*" by European attempts to account for the colonial other (see chapter 3). Since that inaugural moment, Derrida argues, anthropology has been

a European science employing traditional concepts, however much it may struggle against them. Consequently, whether he [*sic*] wants to or not – and *this does not depend on a decision* on his part – the ethnologist accepts into his discourse the premises of ethnocentrism at the very moment when he denounces them. This necessity is irreducible; it is not a historical contingency.... But if no one can escape this necessity... this does not mean that all the ways of giving in to it are of equal pertinence. The quality and fecundity of a discourse are perhaps measured by the critical rigor with which this relation to the history of metaphysics and to inherited concepts is thought.[63]

Derrida's argument raises important questions that I cannot take up here (what are the conditions for knowing how to give in to the necessity of Eurocentrism?). But it provides us with a lens to reconsider Osborn's position, to measure the rigor with which she thought her relation to the history of Mayanism and inherited colonial concepts.

Before Osborn's arrival, the TRDP experts wrote that it would be necessary "to understand the workings of the present [Maya farm] system in full, in order to assess the impact of change, or to know whether or not the Maya, or a proportion of them, will have the skills and adaptability to manage new forms of ownership."[64] The TRDP experts wanted to understand the Maya farm system and whether some "proportion" of the Maya might be capable of owning land.

Capitalism qua development asks anthropology to explain the other qua ethnos in such a way that makes resources available for development (such as that "proportion of them" that may accommodate "new forms of ownership," i.e., capitalist social relations). And this has always been a task specific to anthropology, even when it is not executed by an authorized anthropologist. To continue reading the same TRDP memo, the experts hoped that the anthropologist may plan, "based on her experience and research . . . *a rate of change* which can be easily assimilated by the Maya and mitigate cultural disruption."[65] (After "rate of change," insert "by capitalism qua development.")

Osborn struggled with this assignment more than any other of her time with TRDP. She delayed writing the "Reservations report," burying her draft versions with ink and marginal notes, and completed it only just before departing from Belize, burying it as an appendix to her last TRDP report.[66] Fundamentally the assignment called for her to apply ethnographically derived knowledge of the Maya to find ways to convince them to change the way they manage land, and she wanted to find some way to validate, to preserve, their distinctive indigenous land use patterns. In one long sentence in the report's introduction, she returned to the center of the Maya farm system discourse:

> I will not go into the pros and cons of milpa agriculture but I shall merely state that its practice is not generally suited to the western notion of "private property" and that to radically break the system will result in the disintegration of the Maya[,] who . . . are part of and integrated into the nation, . . . vote and . . . are a contributing part of the economy.[67]

The Maya are good and productive citizens – they vote and contribute to the economy – and therefore their system should be respected. This is a specific and strange way to conceptualize nationhood, but let it pass. For at the same time, Osborn could not avow the status quo of Indian reservations. She too saw them as a colonial anachronism, and probably sensed that they would be eliminated. Thus she made the case for, in her words, "not either for or against reservations but rather to seek a compromise and . . . allow the Ministry of Natural Resources time to assess the situation and also allow the Maya time to adjust in an informed way to possible alternatives to their current system."[68] Note that this is not dissimilar from the position articulated by the Reservation Lands Committee; however, forging this compromise, for Osborn, did not require the views of the Maya. She proposes the compromise herself, one including dereservation, the "strict control of villages," an increase of land tax, and numerous new land use rules – to be promulgated in the capital city of Belmopan.[69]

Caught between the structures of anthropology and capitalism qua development, Osborn responded – within the spirit of Mayanism – by attempting to empirically represent the Maya qua ethnos for positive ends. And though she may have resented it, she did so in order to facilitate their proper development. Notwithstanding her critical reflections on "development" within the project, and her perhaps unfair and untimely departure, her work with TRDP illustrates how the desire to know and speak for the other does not challenge in any fundamental way the settlement between capitalism and development.[70]

Rice and Debt: The Toledo Small Farmer Development Project

Shortly after TRDP's ignoble end, another project, the Toledo Small Farmer Development Project (TSFDP), took its place. Whereas TRDP was one of the last colonial grants from the British, TSFDP represented the entry of Toledo as an object-space in need of development *loans*. The capital came from IFAD, an arm of the UN charged with financing agricultural development.[71] The founding project document outlines the TSFDP's object:

> The predominant farming system practiced in the project area is the milpa system of slash and burn shifting cultivation. Under this system the milpero hand-fells the forest area with a [machete], burns and plants crops.... This system utilizes intensive inputs of labour and extensive use of land. Little inputs are used apart from labour and extensive use of land.... In the past there was a long fallow of about 30 years and a short fallow of about 5–6 years. Due to increasing land pressure there is no longer a long fallow and the short fallow is now the only one in use.[72]

To counter the increasing land pressure, TSFDP proposed an agricultural development program for "improved farming systems" based on "*sedentarized*, improved agriculture, in blocks of 50 acres per family...to gradually achieve substancial [*sic*] production and farm income increases for 1,400 farm households."[73] TSFDP therefore resumed TRDP's work, proposing to overcome rural Toledo's overpopulation and underdevelopment by mechanizing rice, extending microcredit, privatizing the reservations, and settling agriculture, albeit with less commitment to basic research and (in keeping with the times) a stronger emphasis on the market and private property.

The first aim of the TSFDP project was familiar: to "replace the shifting cultivation, which prevailed among small farmers" and thereby improve "the standard of living of a group of small, mostly subsistence farmers now living in isolation, by bringing them into modern agriculture."[74] The TSFDP failed in this respect. In the 1972/73 season, *milpa* rice farmers in Toledo sold 4,654,000 lbs of rice padi to the state marketing board (or BMB; see figure 5.1).[75] Over the four seasons before TRDP arrived (1972–6), the BMB purchased 15 million lbs of rice padi. The majority of this rice was grown by Maya smallholder peasants who planted rice with a dibble stick after using a machete and fire to prepare the land.[76] Thus, when TRDP started in 1978, southern Belize was the rice-basket of Belize, producing more than 75 percent of the rice in the country. Yet after two development projects intended to increase rice production in the south, *milpa* rice production in Toledo – measured as a percentage of national rice production – had all but disappeared.

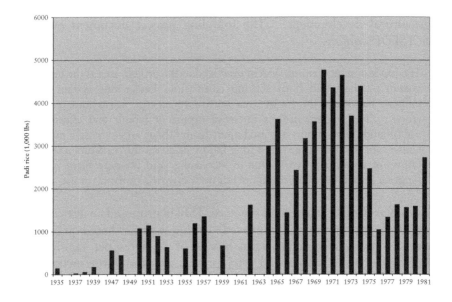

Figure 5.1 Belize Marketing Board rice purchases, 1935–1981 (data: TRDP and BMB)

Over the course of the TSFDP project, the importance of Toledo's *milpa* rice to national production declined steadily (figure 5.2). Not only did the proportion of Toledo *milpa* rice decline relative to other areas, but the absolute quantity of Toledo *milpa* rice production also declined from the 1970s. By 1993, the TSFDP saw some success in conversion to mechanical rice, and *milpa* rice came to make up only 40 percent of Toledo's rice acreage. In 1988, there were over 4,000 acres of *milpa* rice in Toledo, and only 200 acres of mechanized rice. Through the duration of these two projects, *milpa* rice acreage stayed relatively constant, moving between 2,000 to 5,000 acres of land under *milpa* rice. By 1993, more than half the 5,000 acres of land under rice were mechanized. The TRDP and TSFDP facilitated this change by providing basic agronomic research and discerning varieties appropriate for mechanized management. Moreover, TSFDP subsidized the land preparation needed to convert to mechanized rice production: clearing, plowing, harrowing, and harvesting. These contributions were not scale-neutral, since they provided marginally greater gains for the more mechanically savvy and capital-rich farmers.

Yet the trend toward mechanized rice did not last long (figure 5.3). Mechanized rice production in Toledo peaked in the 1993/4 season, the last full year of the TSFDP project. After TSFDP's departure, both acreage and productivity of mechanized rice declined precipitously as

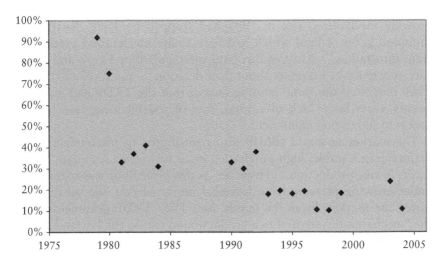

Figure 5.2 Toledo District *milpa* rice production as a percentage of total Belize rice production, 1975–2005 (Data: ESTAP, 2000)

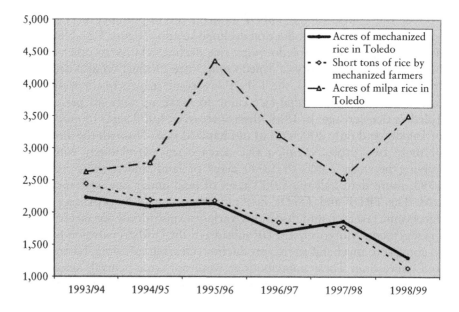

Figure 5.3   The decline of mechanized rice in the Toledo District, 1993–1999 (data: Belize Ministry of Agriculture)

capital-poor Maya farmers got out. Forty-three percent of the farmers growing mechanized rice in 1995 had stopped by 1998. Nearly one thousand acres of land which had been under mechanized production went into fallow.[77] Most of this land surrounds Blue Creek and Agua-cate, where I asked farmers about their decision to quit mechanized rice. They reiterated the same limiting factors that the TRDP had analyzed twenty years later: lack of capital, fear of indebtedness, and lack of access to harvesting equipment.

The marketing board (BMB) also contributed to mechanization by maintaining a stable, high purchasing price for rice to stimulate invest-ments in rice production.[78] However, as this price floor was maintained under growing pressure from low-cost imported rice, the subsidy dis-torted the market. From the outset, as a 1977 TRDP document notes, the "mechanized system of rice production in use on the lowland area is beyond the resources of the average milpa farmer and has yet to prove [to be] financially attractive."[79] Yet the rice marketing board was already purchasing domestic rice at a price that was higher than the import price: Bz $.38/lb in 1977 when imported rice was available for

$.32/lb.[80] As early as 1978, an internal report on the potential consequences of the TRDP project explained that rice "cultivation is already widespread,"[81] and that permanent cropping was prevented by skepticism among farmers due to "the failure of government rice cultivation.... The cost of mechanical farming engenders fear of the need for massive indebtedness."[82] Reviewing the progress with the agronomic experiments at a time when the "whole of [TRDP's] Blue Creek farm was planted to rice," the program director writes:

> Production costs [for rice] are on the high side and efforts to reduce them in the future will be made, particularly through a reduction in the use of heavy equipment. Harvesting once again has been a major problem, with the unreliability of the Agriculture Department. TRDP cannot be expected to invest in a large acreage of rice if it cannot harvest it adequately.[83]

If the TRDP could not be expected to invest in rice mechanization without reliable and adequate harvesting technology, why should the Maya? A project review in 1981 clarifies: "Full [rice] mechanization with [mechanical combines] will be extremely expensive.... The local population have become used to hiring machines at hopelessly uneconomic prices and would probably object to paying the true price."[84] Maya farmers lacked sufficient capital to buy into the mechanized-rice model; the agronomic tests showed that rice could not be produced efficiently (measured against the import price); and for poor peasants the risk of possibly striking it rich with rice were easily outweighed with the time-tested, low-risk approach of growing corn in the *milpa*.[85] So even before the TRDP experts' houses were completed in Blue Creek, one of the project's basic premises – that Maya rice farming could be mechanized economically – was recognized to be wrong.

Indebtedness and the Reservation Lands

The TRDP and TSFDP were organized around a dilemma: increasing mechanized rice production required that local farmers could finance the land preparation and harvesting, yet the farmers had no collateral against which to borrow. The only means of production at their disposal – apart from their own labor – consisted of their collective, usufructory use of the reservations. This is a classic agrarian dilemma, and the project experts responded in the classic fashion: proposing, without any input from the local farmers, to subdivide and demarcate the reservations (common lands) into private parcels, which would be titled to the male "head of household." When this scheme was introduced to the community leaders, not everyone was enthusiastic. Boundary-demarcation meant setting lines in place on the landscape, individualizing the social relations of land. The reservation framework had provided a means to keep those social relations fluid. This marked the first time that reservations have been subdivided in southern Belize, and village elders voiced concerns with departing from tradition (figure 5.4).[86]

At the same time that the reservation lands were divided, farmers turned over their land papers to the Development Finance Cooperation (or DFC), a national development bank. The farmers never actually received any cash; using their new 50-acre blocks as collateral, they received the loans in the form of inputs for rice, citrus, and cattle farming. But problems quickly emerged. The state never delivered on its promises of support, particularly with subsidized mechanization and transportation to markets. Without assistance in transport and marketing, the farmers could not sell their crops profitably. Consequently, most

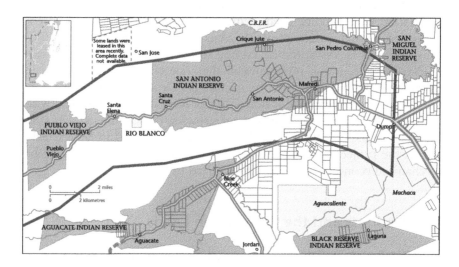

Figure 5.4   The division of Indian Reservation lands, 1993–2003

Cartographer: Eric Leinberger, 2005. The dark grey areas on the map indicate lands that are legally registered as Indian Reservations. The numerous rectangular blocks indicate areas where the Government has subdivided land, usually as leases. The leases around and between the villages of Aguacate and Blue Creek were formed and divided by the TSFDP project.

of the farmers who joined the projects were left in debt. The DFC demanded that they pay back the full value of the loans, of course, and held their land papers as collateral. The farmers received their first letters from the DFC warning of potential foreclosure shortly before Julian Cho's 1995 visit to their communities.

The farmers feared that the DFC would put up their now-divided lands for sale, but Toledo has a relatively inactive land market and few landowners with capital to buy land. One exception is a group of expatriate Mennonite farmers who settled on land near Blue Creek twenty years earlier. The migration of the Mennonites to Blue Creek is a recent chapter in a broader story of their migration to Belize, which followed an agreement signed by the government in 1957 guaranteeing the Mennonites rights in Belize (including their own churches and schools, exemption of immigration deposits, etc.) in exchange for paying all the expenses of "establishing their settlement" and bringing into the colony "investment in cash and kind amounting to five hundred thousand dollars...in British Honduras currency."[87] Over time these settlers have become an important economic factor in Belize's agro-food system, having come to dominate grain and chicken production. As their community has grown, they have spread to new areas, purchasing land throughout rural Belize. When Mennonite settlers came to Blue Creek, they did not pay for land, but offered to assist the community with agricultural development.[88]

In one of those quirks of social change so commonly inflected by the sign "development," the major benefactor of the TRDP and TSFDP projects was this group of Mennonite farmers who migrated into Blue Creek to build up their church in an almost exclusively Catholic community. The Mennonites were not dependent upon the Ministry of Agriculture to provide assistance with tilling, planting, harvesting, and hauling. Thus, the Mennonite community saw opportunity in these projects: with access to capital and machinery, and the capacity to carry out mechanical operations and repairs, they were able to take advantage of the push for mechanized rice. Over time many of the local Maya became de facto contract farmers, taking out loans from the DFC and then contracting with Mennonites for mechanical assistance, field preparation, and harvest. Rice is a time-sensitive crop that must be harvested at the proper time. Consequently Maya farmers came to recognize the benefits of attending the Mennonite church, since being articulated into the Mennonite social network increased their chance of planting and harvesting at the right time. Before the TRDP and TSFDP arrived in these communities, nearly everyone professed the Catholic faith. But since the 1980s, a threefold division has emerged in Blue Creek and Aguacate: those who were close to the Mennonites, those who never joined, and those who joined but were either kicked out or left because of infractions of the strict social rules (especially one – no drinking). Only those who stayed in the fold are likely to be growing rice today. The TRDP and TSFDP thus inadvertently contributed to the rise of "Mennonite" rice – mechanized and capital-intensive.

Today, rice production remains uneconomical in southern Belize, insofar as it cannot be grown and sold profitably at prices that are competitive with imported rice. The government therefore maintains import tariffs, allowing the price of rice to increase during these projects to induce local farmers to take the risk of investing in rice production. Rice subsidy is distributed through price supports and basic agronomic research, but not (as was expected) by subsidizing the mechanical costs of rice production, i.e., the total cost of preparing the fields, harvesting, and delivering the rice to the mill. Naturally the Government of Belize is under pressure today to reduce the subsidy given to domestic rice production. The Ministry of Agriculture recognizes that the gains have been reaped by capital-intensive Mennonite farmers, and therefore the political pressure to maintain the subsidy is declining. Is it any wonder that some in the Ministry of Agriculture have encouraged Toledo's Maya farmers to market their product in the US as organic, fair trade "rainforest rice"?

Developing Resistance

In 2002, I happened to visit the home of an old friend who had been elected to serve as Aguacate's community researcher for the *Maya Atlas* project (see chapter 6). He remembered me through the *Atlas* and my collaboration with Cho, and we spoke of these projects, the unpaid debts, and the land papers. Inspired to do something to change the situation, we organized a series of meetings during which we (I mean the indebted farmers of the village and myself) articulated a critique of the situation and evaluated our options. The first step, put forward by the farmers, was to write a letter to the Prime Minister. The core of that letter, signed by all the indebted farmers of Blue Creek and Aguacate, read:

> We are writing to request your intervention into a problem that has caused us great difficulties for almost a decade. We, the undersigned, are the farmers of Aguacate and Blue Creek villages who were left indebted to the

DFC after the failed Toledo Small Farmers' Development Program (TSFDP).... [W]e, the farmers, should not be held accountable for the debts to DFC, because we took our loans as part of a development program that promised markets and technical assistance which never materialized; ... [A] solution to this problem can only be found by bringing DFC, IFAD, and the Government of Belize together to negotiate a just settlement. We urge you to facilitate a resolution to the problem. Our livelihoods depend on it.[89]

Over the course of the next three years, I collaborated with the farmers in these communities in building a small but disciplined movement aimed at debt cancellation and land papers. The matter of the campaign was the stuff of most social movements: meetings, alliance-building, fundraising, lobbying, press work, confrontation. At its core the movement was comprised of a series of representations of the "debt problem," targeting the Prime Minister, the DFC, and IFAD. The farmers sent six letters to the Prime Minister, two delegations to the capital city, and were featured in an award-winning television report and in the leading newspapers of Belize.

The dénouement of the struggle arrived in April, 2006. Under pressure from the Maya movement to address the land tenure issue, the government's loss in the case before the IACHR, and poverty in rural Toledo, the Prime Minister made a trip to the south to tour Maya communities. Adroit lobbying by Cristina Coc, a young Maya activist and the Director of the Julian Cho Society, put Blue Creek on the tour map. Coc then worked with the local farmers to stage a powerful presentation about the failed projects, the debt, and the lands, and on that afternoon in Blue Creek the Prime Minister faced a highly disciplined, focused, and articulate group of angry and desperate farmers. After Cristina Coc delivered an impassioned concluding speech, the Prime Minister promised to have the debts cancelled, the land returned. One month later, the farmers received their land papers and saw their debts stamped "cancelled."

The movement for land rights and debt cancellation was not the only site of resistance that emerged from the ruins of the TRDP and TSFDP development projects. The ruins themselves became material for an experiment in postcolonial pedagogy when activist Maya teachers from the Cayo District of Belize convinced the government to allow them to take over the abandoned TRDP buildings in order to establish a new "Maya high school." The result of their labor, *Tumul K'in* (Mopan Maya for "new day"), is an institution that provides "an intercultural education that reflects the values and aspirations of Maya people."

Today *Tumul K'in* provides secondary education for about fifty "underprivileged Maya youth who have not had access to a secondary education or have been failed by the conventional school setting."[90] The students take courses in conventional subjects, but also agriculture and agro-processing, ecotourism, and natural resources management. Each student is responsible for contributing voluntary labor to the school, and with this contribution, the lands around the TRDP site are today planted with corn, beans, vegetable gardens, and classrooms.

Back in 1983, when one TRDP project evaluator saw the considerable funds that the British had sunk into the project, he commented on how unfortunate it would be "if the [research] station is allowed to fall into a state of dilapidation and if its facilities are not put to good use."[91] Ironically, it was perhaps only because the research station fell into dilapidation that it could be seized for radical ends. Like the farmers who struggled to win their land papers, the families, students, and leaders building *Tumul K'in* are planting seeds of a new day in the ruins of development.[92]

Conclusion

In the last section of the first volume of *Capital*, devoted to "the so-called primitive accumulation," Marx draws a distinction between the ordinary accumulation of surplus value through the labor process and the accumulation that stems from the dispossession of resources from subaltern groups. The later form – dispossession outside of the "normal" circuits of exchange – he calls primitive accumulation:

> The capitalist system presupposes the complete separation of the laborers from all property in the means by which they can realize their labor. As soon as capitalist production is on its own legs, it...reproduces [this separation] on a continually extending scale. The process, therefore, that clears the way for the capitalist system, can be none other than the process which takes away from the laborer the possession of his means of production; a process that transforms, on the one hand, the social means of existence and of production into capital; and on the other, the immediate producers into wage-laborers. The so-called primitive accumulation, therefore, is nothing else than the historical process of *divorcing the producer from the means of production*.[93]

There is nothing ancient about primitive accumulation; it is an iterative process that facilitates new rounds of capital circulation and accumulation.[94] In his illustrations, Marx emphasized the direct, even kinetic force of producing dependent proletarians: the closing of the commons; slavery; theft of land through colonization; taxation, etc. Yet today primitive accumulation often appears bloodless, since it is through finance's surgical work that debt pries land from peasant hands.[95] In Blue Creek and Aguacate the commons were subdivided to provide collateral for development micro-loans, and farmers nearly lost their land for them. Lack of credit and technical training meant that the Maya were unable to realize the potential benefits of increased rice production. Spivak offers a lapidary critique of development projects that push women's micro-lending, calling it "credit-baiting without infrastructural involvement" that "opens the poorest rural women into direct commercial exploitation by the international commercial sector."[96] And what happens when the poorest of the poor default on micro-loans?

At their essence, these two development projects combined credit-baiting without infrastructural involvement with the incorporation of commonly managed lands into capitalist social relations. This formula is not uncommon today. David Harvey rightly uses this to illustrate how primitive accumulation carries on today, through the "displacement of peasant populations and the formation of a landless proletariat... [where] many formerly common property resources... have been privatized... and brought within the capitalist logic of accumulation."[97] In Blue Creek it was not enough for the land to be parceled to unencumber the peasant from the means of production. Dividing the commons converted the individual peasant households from petty producers who owned land collectively into farmers who owned private property. This shift was tied – legally, agronomically, and financially – to a shift towards the mechanical production of rice. The development experts saw that the nascent commercial rice farmers of Blue Creek and Aguacate would need capital, and that they had none. They recognized that the only means of production at the farmers' disposal – apart from their household labor – consisted of their collective, usufructory use of the reservation lands. So they divided the reservation lands and gave "micro-loans," causing profound and unexpected shifts in local social relations: of religion (the Mennonite dynamic), of the means of production (privatization), and of gender (the land is now titled to men).[98]

And yet these development projects did not fully "liberate" the peasant – not *yet*. Common property was divided, but primitive accumulation was checked by peasant resistance. As the literature on agrarian change teaches us, the diverse forms of agrarian resistance fail more often than not.[99] Of course, "failure" is a relative term. I have claimed that the victory in Blue Creek and Aguacate was an "unmitigated success."[100] And in one crucial sense, it was: the farmers won what they demanded. From their initial meeting with Julian Cho, they called for the cancellation of their debts and the return of their land papers. These two goals were achieved through their mobilization – their successful articulation of claims, texts, bodies, voices, and demonstrations.[101] Primitive accumulation was suspended; today the farmers own their own land, clear of debt.

But this only underscores the depth of the hegemony of capitalism qua development. For if we survey the recent struggle against the broader terrain of the social relations of agricultural production wrought through the Maya encounter with colonial capitalism, what stands out from this episode is that *in victory* the farmers themselves have privatized commonly held lands. Remember that before these development projects got underway, these lands were within an "Indian Reservation," and legally owned by the colonial state, and yet locally held and managed as *commons*. Today these lands are mainly the private property of marginal, capital-poor farmers. I add that the ownership of the land has been transferred exclusively to *male* "heads of household." Although the farmers I have worked with in this campaign have expressed their anxiety about the intensification of interclass rivalries within their communities – through the creation of a market in land – there is little discernable energy to reconstitute these lands as a new commons, one distinct from the Indian Reservation and the state, under the hegemony of a local, indigenous council. In sum, success has been accomplished by way of completing the privatization of the commons. As commons, the reservations exist as social relations, the result of a colonial dialectic of settling and unsettling.[102] More: remember that the initial, political purpose of the TRDP was to cause "the Maya Indians...to identify themselves more permanently with the rest of the country."[103] The path of resistance open to the farmers ran through the privatization of the commons and their self-representation as beleaguered *national citizens*, poor Belizean farmers, to the office of the Prime Minister.

These experiences in Blue Creek and Aguacate bear out an aporia of agrarian resistance carried out against the hegemony of capitalism qua development. For marginal, indebted peasants, faced with the threat of landlessness, a campaign for landownership is something that one cannot not want. And yet the routes open to such campaigns are paved by relations of state and society, law and property, forged in the crucible of colonial capitalism. In this case the privatization of the commons was won by appeals to justice for indigenous people, both *in consequence of* and *in the name of* development. Regardless of whether the peasant farmers in question happen to qualify as "indigenous" before the law, the challenge of turning a global form of power against itself exceeds their capacities. And peasant livelihoods are unsettled by development.

Though the metaphysics of settling may be definitive for capitalism qua development, farmers are not always convinced. Some are still planting rice in Blue Creek, but in the wake of the flood of DFC debt, longstanding patterns are rewoven on riverbanks as farmers choose low-risk strategies. Rice is planted this year, but less than in the 1990s, and little of it mechanized. The farmers that I collaborated with in the campaign for debt relief have returned to the low-risk, low-pay livelihood strategies that have held them to this place, planting maize in the limestone-rich soils along the mid-reaches of the Moho River.

Notes

1 This is the lesson of the economistic literature on the so-called "tragedy of the commons." There is a substantial counter-literature on the socionatural production and management of the commons: see the Digital Library of the Commons (www.dlc.dlib.indiana.edu/contentguidelines.html). An earlier version of this chapter was presented at the 2004 meeting of the International Association for the Study of Common Property in Oaxaca, México.

2 This chapter is based upon fieldwork (including here archival research, interviews, and participant observation) conducted between November 1995 and May 2006. I have visited Blue Creek and Aguacate perhaps thirty times, often staying for several days, for a total of four or five months. Most of my time was spent hanging out with farmers, talking about agriculture, the weather, religion, and debts (in Q'eqchi' and English). I also conducted 40 interviews of varying degrees of formality. I initiated a community survey but ceased when I sensed that it was distorting the interactions that I was having in the village. Before the files were burned to make space for a dorm room, I sorted and read the archive left by the TRDP and TSFDP: I refer to this collection here as the "Tumul K'in archive" (I thank Angel Tzek for access). What I have learned in these two communities comes as a consequence of my involvement in the campaign to cancel the DFC debts and win the land papers.

3 Julian agreed. He went on to make inquiries into their case but – occupied with the conflict over the Columbia River forest, a lawsuit against the government, and the *Atlas* – he made no progress on their problems. The fate of Blue Creek and Aguacate became Exhibit B in Cho's speeches in other Maya villages, used to illustrate the fate of their communities if the Indian reservations were parcelized.

4   Speaking about the prospect for "fieldwork without transcoding," Spivak explains:

> You are a person who is clearly not a subaltern person, who has moved into a group which clearly is subaltern with no kind of mobility. And you are earning trust so that you can do whatever it is that you are there to do.... That patient effort to learn without the goal of transmitting that learning to others like me...can be described...as fieldwork....My goal is not to produce well-written texts about those experiences. If that were so, then I would not be able to learn because my energies would be focused toward digesting the material for production. It's as simple as that....If your energies are focused toward that, you are constantly processing, and you are processing it into what you already know. You're not learning something. So this is why I say that you should perhaps call it fieldwork, because "learning from below" is too pious sounding. (Sharpe and Spivak 2002: 619–20)

5   Spivak (1993) proposes reading "fieldwork" catachrestically as a practiced "hanging out": there is "the subterranean historical thing in my work, [which is] the exact opposite of the other project of 'hanging out' – *Mitda-sein* as *Mitwegsein* – hanging out so that you can catch the ontic where it is not, the ontico-ontological difference as *différance*...first of all, there is the subterranean historical project, then the surreptitious unlearning." Spivak addresses here the question of whether and how it would be possible to think an ethic outside of the inherited tradition of Western Christian-ethical thought. She proposed (in the ellipses of this quotation) that it should be "possible to imagine [such thought] – because, after all, if one thinks that the ethical possibility is merely another way of saying how one is human...then it should not be...necessary to posit a degree of acculturation in order to be merely ethical." This marks one of the points of distinction between Said and Spivak. Spivak writes that Said "was amused by my on-the-ground political commitments that had to be different from his, for they were "post'-colonial....[W]hen Edward would ask, 'Gayatri, what do you do when you go to those villages?' I would give the usual answer, 'Hang out' (*Mitwegsein*, suspend previous training in order to train yourself, you know). The answer was not satisfactory" (2006: 524).

6   Any sort of fieldwork always entails a capricious and not unaccidental weaving of social texts. This describes the epistemological challenge of "situated knowledge" in the geo-field. But these socio-geographical entries-and-limits do not in themselves constitute the project. Merely accounting for exactly how one gets "in" to the field – even through the most rigorous Gramscian encyclopedia of traces – would not be sufficient to explain how one *thought* in that space. The work of taking account of fieldwork is therefore both necessary but always insufficient – to itself as such, but also to the basic problem of defining the boundaries of fieldwork-knowledge – because the play of questioning, thought and writing are still to come. Awareness of "context" (unequal class and power relations, unevenly

traversed geographies, different languages and training) cannot account for the particular *questions* that are raised, nor the power of the there-here distinction. These are always inherited from prior discourses and struggles and reiterated through fieldwork. This inherited play solicits the postcolonial fieldworker and underscores the importance of reading.

7 Don Owen-Lewis correctly called the TRDP project compound "a settlement, settled colony." When Owen-Lewis was asked by the ODA for advice on the project, he opposed the plans for locating the TRDP station in Blue Creek: "I said, 'Don't! You're going to put up a white elephant up there and nobody's going to use it. They [Maya farmers] don't like it. It's too far away.... [P]ut me down on your books as saying that I totally disagree with this thing.' ... [T]hey had about four, five air conditioned, senior staff houses; gardens; they had air-conditioned offices; they had a lovely workshop; they had a bunch of stuff there, and it all went back to bush."

8 Anon., 1977. "Rural Development, Toledo Province: Belize." Barbados: British Development Division in the Caribbean. AB, MC 3667, pp. 1–2.

9 On narratives of capitalist space-economy, see Barnes (1996).

10 Anon., 1977. "Rural Development, Toledo Province: Belize." Barbados: British Development Division in the Caribbean. AB, MC 3667, pp. 1–2. These words are repeated elsewhere in TRDP documents without attribution: cf. TRDP project head J. S. Briggs's 1980 "End of tour report" (Blue Creek: Tumul K'in, p. 1). Development texts – concept papers, proposals, reports – often reiterate basic claims, without attribution, such that initial ideas become goals, goals become a rationale, and a rationale becomes the final report.

11 Anon., 1977. "Rural Development, Toledo Province: Belize." Barbados: British Development Division in the Caribbean. AB, MC 3667, p. 2. Here is the same sentiment from a key project document written three years later, in 1980: "Government policy has stressed the development of agriculture, especially rice cultivation, as a means of *drawing the Toledo District closer to the aims of national development.*" Crapper, D. B.; Harrison, J.; Waddell, B. L., 1980. "Toledo Rural Development Project: A review." Barbados: British Development Division in the Caribbean, TRDP/Tumul K'in, p. 6.

12 Ibid., p. 6.

13 Thomson, B. P. 1977. "Toledo Agricultural Project: Economic Assessment." Barbados: British Development Division in the Caribbean, p. 2.

14 Crapper, D. B.; Harrison, J.; Waddell, B. L., 1980. "Toledo Rural Development Project: A review." Barbados: British Development Division in the Caribbean, TRDP/Tumul K'in, p. 8.

15 Johnson, I. M. 1981. Report on a visit to Belize. Bedford, UK, ODA, p. 12.

16 Crapper, D. B.; Harrison, J.; Waddell, B. L., 1980. "Toledo Rural Development Project: A review." Barbados: British Development Division in the Caribbean, TRDP/Tumul K'in, pp. 6–7.

17 Van Ausdal (2001).
18 Comparing data from the 1973–74 agricultural census and the TRDP Land Pressure Survey of 1984 for the seven rural Maya communities included in both surveys. These seven include mainly communities in what TRDP called its "impact area."
19 Like many former British colonies, Belize has a Westminster-based political system based on multiparty elections, common law, and universal suffrage, with elections oriented around two major parties: the PUP and UDP. The PUP evolved from labor-union roots in the 1950s to mobilize the anti-colonial movement. The UDP formed from a consolidation of several anti-PUP parties in 1973. These two parties have since dominated the political environment, when the PUP negotiated the transition to political independence.
20 R. Holloway, 1976. "Social Survey of the Area Around Mafredi: Toledo, Belize." Barbados: British Development Division in the Caribbean, pp. 12–13.
21 "It was stated by both the Minister of Natural Resources and his adviser...that, whilst it was hoped the Maya would share in the anticipated prosperity the development would stimulate, the ultimate aim was to maximize use of land, which the 'milpa' system hinders" (Crapper, D. B.; Harrison, J.; Waddell, B. L., 1980. "Toledo Rural Development Project: A review." Barbados: British Development Division in the Caribbean, TRDP/Tumul K'in, p. 6).
22 R. Holloway, 1976. "Social Survey of the Area Around Mafredi: Toledo, Belize." Barbados: British Development Division in the Caribbean, p. 2.
23 Abstracting A. C. S. Wright, Holloway argues that the creation of the Indian reservations by the colonial government "set up an artificial situation in which the Indians are *financially protected*, but given population and migration pressure, are held to an amount of land that is too small" (Holloway 1976: 7, my italics). It would be impossible to show that the Mayas have been financially protected.
24 Ibid., p. 3.
25 Ibid., p. 15.
26 The push for mechanized rice in Toledo arguably dates to 1968, when the Dump-Mafredi mechanized rice trials were first established, but the key turning point came in the 1977/78 season with the beginning of TRDP and the construction of Big Falls rice mill. Yet aspirations for rice development have long defined the Toledo District in the imagination of Belize's "national development." Intensive rice cultivation in Belize can be traced to the late 1800s, when Toledo was the center of experimental rice planting. In 1884 the state counted only 1 acre of rice planted in the Toledo District; by 1903 this had grown to 96 (at a time when there were only a few hundred farmers in Toledo). Rice's agronomic development in Toledo was conducted entirely by Maya, Garifuna, and East Indian peasants,

planting rice with dibble sticks on lands that they did not own. State involvement in Toledo's rice production was minimal until 1933, when the colony's Rice Experimental Station (today home to Toledo's Ministry of Agriculture) was established in the heart of Rancho, 5 miles from the sea on Toledo's only paved road (the Agriculture Department was only formed in 1928; the first Agriculture Officer, H. P. Smart, was appointed by the colonial government the same year). Domestic agricultural cultivation, particularly for rice and beans, intensified during World War II as food prices increased. The colonial state made only modest investments to stimulate this trend (e.g., in 1954, a public rice development scheme was created 24 miles northwest of Belize City with a CD&W grant). Today Belize still imports most of the food that is consumed (see Wilk 2006). The value of agricultural exports only surpassed the value of forestry exports in 1960. Belize would earn more from agriculture than any other sector only until the 1990s, when the rise of tourism and services (as well as remittances from the US and black-market exchange) displaced agriculture from the core of economic strategy.

27   Anon., 1976. "Toledo Rice: Civil Engineering Aspects." Barbados: British Development Division in the Caribbean, AB, MC 3667, p. 8.

28   The local historiography reflects the colonial trope of settling. Consider Blue Creek, home to the TRDP "ruins." The local histories that I learned tell different stories of precolonial roots, settlement, the founders of the village. For instance, the authors of the Blue Creek history in *Maya Atlas* make no mention of founding: "from 1925 to 1950, few families lived in Rio Blanco," as the village was known. In 1950, a few families "came from Aguacate and San Antonio. . . . They named the place Blue Creek because of the beauty of the river and hills. Clear water was abundant for washing and fishing; there were hillsides for hunting; and fertile soil for plantations. Today people in Blue Creek speak both Mopan and Ke'kchi Maya because of the earlier settlers" (TMCC and TAA 1997: 57). In this view, Blue Creek was an *alkilo* where farmers would stay during harvest season of the *matahambre* maize, or staying while fishing and hunting near the river. In this case the question of "permanent settlement" becomes especially difficult to pin down. Perhaps this explains why Blue Creek is shown on a 1937 map, before the return of these "few families" (Grant, A. 1937. "Sketch map showing Guatemala Villages from which the Indian immigrants are derived." AB Map 1B). Yet another "History of Blue Creek" is told in a text of that name, based on oral histories collected by one member of the village. It attributes the founding of Blue Creek to a specific event: "In the 1800s three Kekchi families left Guatemala and started a community known as Rio Blanco. . . . Their families originally came from San Pedro Carcha, Coban, Guatemala; they settled in Rio Blanco" (Anon., ca. 1999. "The history of Blue Creek village"). There is no reason to settle these variations into one theme.

29 In an internal critique of TRDP written by the project director in 1980, "serious delays with housing" is the first of three major problems discussed (Briggs, J. S. 1980. End of tour report. Blue Creek: TRDP/Tumul K'in, p. 13).

30 TRDP was funded by an ODA grant as a kind of farewell gift to the colony.

31 Sykes, C. B. 1981. Draft progress report – October 1980 to September 1981 and proposal for future work. Blue Creek: TRDP, p. 3.

32 The development agents always felt that they were short of time: "our concern with time has been all important – the urgent need to get staff housing completed, to get some land developed on the major soil types so that the agronomic work could forge ahead, and the overall sense of hurry that alternative farm systems are needed without delay" (Sykes, C. B. 1981. Draft progress report – October 1980 to September 1981 and proposal for future work. Blue Creek: TRDP, p. 3).

33 BDD, 1982. Toledo Research and Development Project: 1983 Annual Review. Bridgetown, Barbados: Overseas Development Administration. AB, MC 1738, p. 16.

34 Wright (1996a).

35 Ministry of Natural Resources, 1981. Letter to A. M. Archbold, BDD. Belmopan, Belize, Ministry of Natural Resources, pp. 2–3.

36 Ibid., p. 4.

37 Sykes, C. B. 1981. Draft progress report – October 1980 to September 1981 and proposal for future work. Blue Creek: TRDP, p. 2.

38 Ministry of Natural Resources, 1981. Letter to A. M. Archbold (Tumul K'in), p. 4.

39 Understaffing was a major concern because the TRDP could not recruit, train, and maintain their "counterparts" – largely because there was no means or place for non-local Belizean staff to live in Blue Creek. This problem was identified in an internal critique of TRDP written by the project director in 1980, but never addressed (Briggs, J. S. 1980. End of tour report. Blue Creek: Tumul K'in, p. 13). Naturally, the TRDP experts blamed the modest capacity of the new government for this problem. But who bears responsibility for the inheritance of the colonial state, or for British-funded development plans that are badly conceived?

40 BDD, 1982. Toledo Research and Development Project: 1983 Annual Review. Bridgetown, Barbados: Overseas Development Administration. AB, MC 1738, pp. 24–6.

41 Ibid., p. 20.

42 Spivak (1988).

43 TRDP, 1982. TRDP Final Proceedings workshop: delegate register. Belmopan, Belize. AB, MC 2842.

44 To be clear, my critique is not an appeal for mere consideration of Maya views. The desire to develop the other can accommodate liberal-multicultural norms of cultural representation.

45 McCann, G. 1982. Report on visits to CATIE, Costa Rica and TRDP, Belize. London: ODA, p. 12.

46 R. Wilk and Mac Chapin, 1990. *"The Maya." Ethnic Minorities in Belize: Mopan, Kekchi, and Garifuna.* Belize City: SPEAR, p. 53.

47 Anon., 1983. Economic constraints to the development of upland agriculture in Toledo. Blue Creek. TRDP, p. 5.

48 When I first entered the TRDP's old record room the bat guano was thick, the mold considerable. I spent two weeks sorting through these files and compiling notes, before a flood and other concerns drew me away. By the time I returned the following year, the files were ashes – burned to make room for a student's dormitory.

49 Preston, D. 1978. "Social factors and the Toledo Rice Development Project." TRDP: April 21, 1978. AB, MC 3658.

50 Wright (1996a). See also Wright (1996b). The latter is also useful for reconstructing the history of interventions since the 1950s and also for understanding Wright's later views on Maya agriculture.

51 To repeat, today the hegemonic answer to the question "what constitutes development?" is capitalism qua development.

52 Cf. Said (2002 [1988]).

53 Osborn had devoted most of her life to work in Colombia, which she mentions in at least one letter. She was born around 1935. She moved to Colombia in 1958 when she was in her early twenties. She is perhaps best known as an ethnographer of the U'wa, work that she conducted in the 1970s and 1980s. Her doctoral dissertation, "Mythology and Social Structure Among the U'wa of Colombia," was defended at Oxford in 1982, shortly after she left the Toledo and the TRDP project, which suggests that she may have written her dissertation in Blue Creek. In a letter written in 1980 she explained that she "specialized in Chibcha speaking people of Colombia. Working in Belize is a temporary break very necessary to give distance to my own area and work." Osborn died in 1988.

54 Crapper, D. B.; Harrison, J.; Waddell, B. L., 1980. "Toledo Rural Development Project: A review." Barbados: British Development Division in the Caribbean, TRDP/Tumul K'in, p. 17.

55 I found a letter that she wrote to the alcalde about a villager, who worked for TRDP, who owned pigs that came onto the TRDP grounds and did some damage. A perfect example of how the dirty work of social life was turned over to Osborn – so that the male agronomists and economists could Develop.

56 Osborn, A. 1980. Letter to M. Howard. Tumul K'in. Though Osborn was not the "gender person" on the project, as the only woman among the TRDP development experts, she was assigned all "social" analyses (rigorously distinguished from the economic affairs) as well as the project's relations with the local communities (but not the state). In the terms of the time, this was "Women in Development."

57   Osborn, A. 1980. Letter to Dr. Carter. Tumul K'in.

58   Osborn, A. 1980. Letter to S. Dreyfus-Gamelon. Tumul K'in.

59   Osborn, A. 1981. Letter to Hazel. Tumul K'in. In the letter, Osborn writes that the biggest problem facing TRDP is that it is not clear "what kind of project it is, e.g. agricultural or rural development (and if the latter [development] is it really possible to do it!!)."

60   Osborn, A. 1981. Letter to Hazel, March 21, 1981. Tumul K'in.

61   Chatterjee (2001: 17). Chatterjee does not elaborate on the nature of the "relation of power involved in the very conception of the autonomy of cultures," but we can grasp its basic qualities by the contours of his critique of nationalism. Ethnography names a mode of writing that describes/produces the other qua ethnos. It therefore presupposes the "very conception of the autonomy of cultures" that Chatterjee writes of here. It operates through the rigorous marking of observations about cultural traits and the accumulation of empirical facts for the identification of the distinctly other qualities (which the ethnographer is trained to recognize). Writing that does not share these qualities is not ethnographic.

62   See Derrida (1978 [1966]). Though it took 12 years to be published in English in a widely circulated book, Derrida's argument in "Sign, structure, and play" inaugurated the critique known as "post-structuralism" in the USA. I endorse Mowitt's suggestion that "it is a far more compelling thought experiment" to argue that post-structuralism is subordinate to postcolonialism, not the other way around: "Derrida was doubtlessly onto something when, in 'Sign, structure and play...', he noted that the enabling negation of Lévi-Strauss's concept of structure was written into ethnography's failure to think the violence and the limit of the West" (2005: xxix).

63   Derrida (1978 [1966]: 282), my italics. Ismail (2006) makes a stronger version of this claim: "If anthropology could be conceptualized as impossible without the inside/outside distinction, as always representing or speaking for an other, whether understood as located in the cartographic west or outside it, then all anthropology, in a structural sense, is colonialist."

64   Crapper, D. B.; Harrison, J.; Waddell, B. L., 1980. "Toledo Rural Development Project: A review." Barbados: British Development Division in the Caribbean, TRDP/Tumul K'in, p. 7.

65   Ibid.

66   Osborn (1982: 98–111).

67   Ibid., p. 98.

68   Ibid., p. 99.

69   Ibid., pp. 109–10.

70   After Osborn had been let go, the TRDP experts called for collection of "sociological data" because "expertise is needed to examine and to advise on the land tenure situation" (BDD, 1982. Toledo Research and

Development Project: 1983 Annual Review. Bridgetown, Barbados: Overseas Development Administration. AB, MC 1738, p. 31). Writing two decades later, A. C. S. Wright speculated that TRDP "failed because they were lacking a full-time sociologist to advise them." This amounts to criticizing development for its lack of sociocultural analysis, a move we saw earlier vis-à-vis Don Owen-Lewis's failures. This critique is always misplaced. Anthropologists may reply: it is unfair to criticize our discipline by the reading of a marginal ethnographic/development text from one failed project. Fair enough – but isn't anthropology structurally committed to the merits of the singular case study that somehow reveals the broader patterns of Man?

71 IFAD was founded in 1977 as a product of the 1974 World Food Conference that called for a Fund "to finance agricultural development projects primarily for food production in the developing countries" (IFAD, 2006, from: www.ifad.org/governance/ifad/ifad.htm).

72 Proyecto International, 1984. IFAD Preparation Report: Toledo Small Farmers Development Project. AB, MC 1560. Points cited are 067 and 122.

73 Ibid.

74 IFAD, 1994. "Belize: Toledo Small Farmers Development." Accessed 2001 from: www.ifad.org/evaluation/public_html/eksyst/doc/prj/region/pl/belize/r172blbe.htm.

75 Moreover, the BMB buying price in 1972/73 was only 7.5 cents per pound; it was increased to 14 c/lb two years later, and would go higher still (see below). However, the rising price of rice was not enough to stimulate large-scale conversion of mechanized rice production among Maya farmers.

76 The highest producing villages were Big Falls, San Pedro Columbia, Mafredi & Crique Trosa, San Antonio, Indian Creek, Santa Cruz, the Dump, Silver Creek, San Jose, Blue Creek, Laguna, and San Miguel. These 12 villages sold 83 percent of all the rice purchased at the Big Falls rice mill in 1972–6 (Thomson, B. P. 1977. "Toledo Agricultural Project: Economic Assessment." Barbados: British Development Division in the Caribbean, Annex II).

77 Data from ESTAP (2000: 38).

78 Data compiled from three sources: Aldana and Lee (1982: 9); TRDP (ca. 1978); Thomson, B. P. 1977, Annex II. I have been unable to find data for 1954, 1956, 1960–1, and 1963.

79 Thomson, B. P. 1977. "Toledo Agricultural Project: Economic Assessment." Barbados: British Development Division in the Caribbean, p. 5.

80 Ibid., p. 3.

81 Preston, D. 1978. "Social factors and the Toledo Rice Development Project." TRDP: April 21, 1978. AB, MC 3658, pp. 10–11.

82 Ibid.

83 Sykes, C. B. 1981. Draft progress report – October 1980 to September 1981 and proposal for future work. Blue Creek: TRDP, p. 9.

84 Johnson, I. M. 1981. Report on a visit to Belize. Bedford, UK, ODA, p. 15.
85 TRDP found it had little to contribute to local maize production, since yields of their varieties in agronomic trials "remained similar to those obtained by farmers and no introduced factor has made any major difference to yields": Sykes, C. B. 1981. Draft progress report – October 1980 to September 1981 and proposal for future work. Blue Creek: TRDP, p. 17.
86 Subdivision has since been repeated in other Maya communities. The parceling of these Indian reservations was probably illegal. This question remains unresolved by the courts. A team of lawyers has prepared a case against the Government of Belize concerning the parceling of one Indian reservation. I am nominally an "expert advisor" to them.
87 Agreement between the Colonial Government of British Honduras and the Kleingemeinde Mennonite Church, December 16, 1957. AB, MC-3334.
88 The community of Blue Creek debated the Mennonite offer and agreed to let the Mennonites settle. Today they have no written records concerning the agreement. The Mennonites and Mayas are not alone in viewing the Mennonite community as a source for agricultural development of the Maya farm system. One 1973 doctoral dissertation argues that "the Mennonites brought with them sophisticated agrarian skills unknown in an underdeveloped nation of slash-and-burn migrant peasants" (Hall 1973: 65). Therefore, Hall concludes, "the Mennonite agricultural system could be used by the local *milparios* [*sic*] to *develop sedentary commercial farms*" (p. 304, my italics). This is what the Mennonites promised to some of the Mayas of Blue Creek.
89 I helped the farmers to write this letter (dated May 30, 2002), and delivered it to the Prime Minister's office. With this letter we established a protocol that continued for the subsequent letters: the farmers would debate their options in the village meetings until a position was decided upon. I was then instructed to write the letter for the group, then give the letter to the leaders of each community, who would carry it to each house for discussion and signatures.
90 Penados (2006: 6).
91 BDD, 1983. Toledo Research and Development Project: 1983 Annual Review. Bridgetown, Barbados: Overseas Development Administration. AB, MC 1738, p. 22.
92 This point has not been lost on the *Tumul K'in* visionaries: "Tumul Kin Center of Learning ... is ironically and metaphorically located on an abandoned site of well funded development projects that ... produced very few positive results" (Penados 2006). Here I would like to thank my JCS colleague, Filiberto Penados, for sharing his thoughts on the history and future of *Tumul K'in*.
93 Marx (1906: 785–6), my italics.
94 Arguing thus that it is "peculiar to call an ongoing process 'primitive' or original'" (p. 144), David Harvey recently proposed the term

"accumulation by dispossession" to refer to the iterative, cyclical practices of excessive or special appropriation that bring about, through forms of creative destruction, new areas for capitalist investment. The key variable driving accumulation by dispossession, Harvey argues, is *overaccumulation* of capital, that is, the periodic lull in opportunities for capital to be productively invested and a restless "capital surplus" seeks an outlet for application: "accumulation by dispossession . . . releases a set of assets . . . at very low . . . cost. Overaccumulated capital can seize hold of such assets and immediately turn them into profitable use" (Harvey 2003: 149). Much of the overaccumulated capital generated through the high oil prices during the 1970s was recycled/invested through "development" initiatives, through speculative lending (mainly to NICs like Brazil and Mexico) but also through international financial institutions like IFAD, which funded the TSFDP through a loan to the Government of Belize.

95 Marx anticipated that the maturation of global finance would make primitive accumulation less opaque: "with the national debt arose an international credit system, which often conceals one of the sources of primitive accumulation" (Marx 1906: 828).

96 Spivak (1999: 200). Spivak rightly warns that we "cannot dismiss finance capital and notice only the exacerbation of wage labor," or else "credit-baiting can be offered as a 'solution.'" As indeed it is: not only does capitalism "solve" the problem of inequality through development, but finance capital has even been defined – at least by the Nobel Committee – as *peace*. I recommend Davis's blistering critique of the "epistemological fallacies" behind development based upon microcredit and land titling (2006: 179–185).

97 Harvey (2003: 145–6).

98 At the TRDP final proceedings, one development expert reminded the audience that if the project had accomplished its stated aim – "to transform Maya farmers from 'shifting' to 'settled' agriculture" – it would have comprised "a major social and economic transformation" (Brown, M. 1986. Proceedings of the TRDP final workshop, p. 93). TSFDP thus executed a process not completed by TRDP.

99 Cf. Scott (1987).

100 For instance, at the July 2006 meeting of the JCS Board of Directors in Punta Gorda.

101 A "demonstration" is a political performance. In the case of Blue Creek and Aguacate the principle demonstration came through a public confrontation with the Prime Minister: eminently direct, albeit "non-violent."

102 In reply to the neoclassical argument that privatization of land resources is needed to facilitate conservation, the commons literature has rightly illuminated examples of commonly managed lands that are conserved precisely because the socionatural relations are densely wound. But we err if we think that such assemblages derive entirely before colonial

capitalism; the "commons" in Blue Creek and Aguacate are the effect of struggles over colonial rule, colonial/indigenous spaces. The celebration of local forms of common property is amenable to a liberal politics that embraces capitalism qua development. To avoid this, the narrative of "the enclosure of the commons" must attend to the historical-geographical conditions that give rise to particular struggles over commonly managed resources (for which there is no general form or predetermined outcome) and the conditions that tear them asunder. Harvey concludes that the question of whether primitive accumulation is a necessary moment in the movement towards socialism hinges on a distinction drawn between those forms of primitive accumulation that open "a path to expanded reproduction," as opposed to forms that simply destroy paths already opened (Harvey 2003: 163–4). The problem with this argument is that we never know which path is which – not only because of the complexity of political-economic change, and because capital's expanded reproduction carries with it its own destruction, but because there is no ledger to ensure that the fruits of expanded reproduction are distributed justly to those subaltern groups whose resources were dispossessed. Land reform in the communities may well bring about some increase in rice production, but it will not feed those who lose their land.

103   Anon., 1977. "Rural Development, Toledo Province: Belize." Barbados: British Development Division in the Caribbean. AB, MC 3667, pp. 1–2. These exact words are repeated elsewhere in TRDP documents without attribution: see, for instance, TRDP project head J. S. Briggs's 1980 "End of tour report" (Blue Creek: Tumul K'in, p. 1).

6

Finishing the Critique of Cultural Ecology: Reading the *Maya Atlas*

> In the history of colonial invasion maps are always first drawn by the victors, since maps are instruments of conquest. Geography is therefore the art of war but can also be the art of resistance if there is a counter-map and a counter-strategy.
>
> Edward Said (1996: 27)[1]

The early 1990s saw the emergence of the concept of "counter-mapping." The argument was that subaltern social groups could use novel, inexpensive forms of geo-referencing technology to produce maps of their own communities – maps that would help to represent indigenous land claims and conserve natural resources. For its advocates, such mapmaking promised to give subaltern groups a means to articulate subaltern geographies, even to make geography an "art of resistance" and "counter-map[ping]," in Edward Said's felicitous terms.

Once defined as such, the concept caught on quickly. A special issue of the journal *Cultural Survival Quarterly* in 1995 documented scores of counter-mapping projects underway in different parts of the world. Canada, Southeast Asia, the western United States, and Central America were discernable centers of counter-mapping work.[2] The lead editorial argued that such projects could convert indigenous lands with ambiguous title into "territory that its occupants can legitimately defend."[3] As one of the most vocal advocates of counter-mapping, UC-Berkeley geographer Bernard Nietschmann, wrote in that issue:

> Many indigenous peoples worldwide are engaged in the *revolutionary* undertaking of accurately mapping their own land and sea territories using the new geomatic technology of GPS and GIS, coupled to computers, video and film. The approach and objectives are very straightforward: defend the land and sea territories by accurately mapping them using new technology and traditional knowledge; use the maps to show the territories have been long occupied and managed which is why they are still biologically rich; and use the maps to internationally promote political decentralization as a geographic-territorial fact and the best means to conserve the planet's biological and cultural diversity.... [T]his indigenous strategy is almost basically unstoppable. It effectively turns upside down the notions behind the maps that justify the intrusions.... More indigenous territory can be reclaimed and defended by maps than by guns.[4]

At the time Nietschmann's essay was published, I was collaborating with leaders of the Toledo Maya Cultural Council (TMCC) to write a proposal for a "Maya mapping project." We proposed to discern and map the boundaries to all of the Maya communities and the broader Maya historical and cultural land use areas in southern Belize.[5] When funding for our proposal was serendipitously offered by the Government of Luxembourg in early 1996, the TMCC and the Toledo Alcaldes Association (TAA) hired Nietschmann and his team of geographers to come to southern Belize and administer the project. I was invited by the TMCC to assist Nietschmann's team.

An extraordinary example of the possibilities of counter-mapping soon unfolded. One "community researcher" was selected from each of the 42 Maya communities in southern Belize. These researchers gathered for two weeks of intensive training in basic cartographic and community research methods. Over the course of the next month, the researchers conducted surveys and mapped their own communities. Their work resulted in the publication of the *Maya Atlas*, a striking 155-page text comprised of maps, oral statements, photos, and drawings compiled by a team of about sixty Maya leaders. As the inside cover of the *Atlas* explains:

> The *Maya Atlas* was made by the forty-two Ke'kchi and Mopan Maya communities of southern Belize. The maps, text, drawings, photographs and interviews were done by Maya village researchers and cartographers

elected by the communities. In their own words and with their own maps, the Maya describe their land and their life, the threats to their culture and rain forest, and their desire to protect and manage their own Homeland.[6]

As with other counter-mapping projects, the *Maya Atlas* articulates an argument for indigenous rights to a distinct cultural space in a way that would resound with a transnational audience. I use the term "articulate" here decidedly, in Stuart Hall's sense: the *Maya Atlas* not only needed to represent the Mayas of Belize and the geography of their region, but to do so in a way that showed these two things to be essentially *joined*.[7] That this entails the geographic articulation of a space and a culture is evident by the reference – written in the third person by an implicitly non-Maya author – that the *Atlas* represents a desire for "*their own Homeland*." This notion of a Homeland is one of the key tropes in the *Atlas*:

> The ultimate dream of the Mayas is yet to be achieved – the security of tenure to their lands and the creation of a 500,000-acre Maya Homeland.... [T]he Mayas are not seeking the establishment of a separate state, but merely secure land tenure, a fair share of the Mayas' patrimony, and a meaningful sustainable relationship with Mother Nature.[8]

When the *Maya Atlas* was released in 1997, over five hundred people came from southern Belize to celebrate in the capital city of Belmopan. Maya leaders gave speeches urging the government to transfer land rights in the south to the Maya. Julian Cho, the Chairman of TMCC, handed a copy of the *Maya Atlas* to Said Musa, soon to become Belize's Prime Minister. Diego Bol, one of the TMCC's founders and coordinator of the *Atlas* project, argued:

> The *Maya Atlas*...is our tool to show our existence, a weapon to press for our legal right to a piece of the jewel, our desire to be active participants in the sustainable use of our resources, and our endless cry for respect to Mother Earth.[9]

Two things should strike us immediately about these three statements from Bol and the *Maya Atlas*. First, they suggest what is at stake in reading and thinking through counter-maps. The *Atlas* is not simply the product of an empirical exercise in "stating the facts" about southern Belize through the representation of spatial data. A reading that focuses on its empirical utility misses the point that the *Atlas* reflects an effort to decolonize space. In this sense, counter-mapping is a name for the effort to lay the groundwork for a different kind of livelihood politics. Second, their force derives from the iteration of tropes that are not themselves "indigenous" to southern Belize: sustainable development ("sustainable use of our resources"), nationalism ("Maya Homeland"), international law ("our legal right"), and feminized nature ("Mother Nature," "respect to Mother Earth"). These tropes predate and exceed the *Atlas*, and we should question their value for an anticolonial politics.

There is today a substantial literature on counter-mapping.[10] Much of the literature is concerned with its practice and questions about the effectiveness of indigenous geographies for conservation practices or legal cases. A subset of the literature, where I would locate this chapter, is directly concerned with the politics of counter-mapping. Nancy Lee Peluso has argued that the key questions about the political effects of counter-mapping are

> to what degree new notions of territoriality reflect older ones; how the reinvention of these traditions benefits or works to the detriment of customary practice, law, and resource distribution; and how the intervention of NGOs [and geographers] affects the villagers' access to and control over...resources.[11]

I hope to take these questions up, but also to pose new ones as well, by reading the *Atlas* within a postcolonial purview. The *Atlas* confronts us with aporias of writing postcolonial geographies. In trying to undo colonial geographical knowledges, the effort to write "counter-maps" represents the vanguard of postcolonial geography. Indeed, what defines these projects as *counter*-mapping is their explicit confrontation of the hegemonic geographies which they contend with and rework – that is, to bring into full presence the language of the radical other. In this case, the other in question is not simply a subject (although this is also at stake), but more particularly a way of conceptualizing the worldliness of the world.[12] In short, the *Atlas* is a work of anticolonial geography that seeks to show the reader an other world that is not recognized – to bring before us a world that we have not been able to see.[13] Reading the *Atlas*, we can see how the attempt to write this world fails. This "failure" is worth pondering. The work of turning cartography against colonialism cannot occur outside of, or after, the colonial present. Therefore the failure to bring a radically other world into full presence is nothing to be ashamed of. It is, rather, a productive moment for us to *read* carefully – precisely because it suggests the task at hand for postcolonial geography.[14]

With this in mind, I aim to reread the *Atlas* with two aims. The first is to read its constitutive elements – the photos, essays, drawings, but especially its maps – to and for their limits. The premise of this reading exercise is that we cannot read the *Atlas* as a reflection of an essentially Maya space. To do so would require one to disregard the conditions of the text's production – such as the capital, languages, concepts, authors, and tools drawn from elsewhere. We would also have to ignore the fact that colonial discourses (about nature, Maya culture, and Belize) are necessarily recapitulated in the *Atlas*. Following Spivak's argument that "all...clear-cut nostalgias for lost origins are suspect, especially as grounds for counterhegemonic ideological production,"[15] I approach the *Atlas* not as a pure reflection of "the Maya cultural region," but as a text that seeks precisely to *produce* this space. Rather than mapping a singular, essentially Maya landscape, my reading of the *Atlas* tracks the ways it articulates Maya spatial identities. I ask: what makes such counter-mapping a "revolutionary undertaking," in Nietschmann's sense? What does counter-mapping *counter*, and exactly how? Indeed, how does a counter-map *work*? What are the politics of counter-mapping as spatial-identity production? If it is an "indigenous strategy" as Nietschmann suggests, then what or where is it indigenous *to*? One (perhaps too concise) answer to these questions is to say that the counter-mapping of the *Atlas* entails the worlding of cultural space as national territory. If this is so, we should ask: what sort of worldliness is written here – and *how*?

This question suggests the second aim of my reading: to consider the *Maya Atlas* not through the lens of postcolonial geography, but as a work of postcolonial geography. Counter-mapping names an irresistible anticolonial strategy, one that we cannot *not* desire in the experience of thinking through the colonial present. The *Atlas* reflects a strategy for decolonizing space, one that articulates an indigenous-nationalist approach to nature and culture with a project of territorialization. Insofar as postcolonial theory argues that nationalism is insufficient to the task of overcoming colonialism, then we may read within the complications around the Maya nationalism of the *Atlas* a postcolonial critique of Mayanism.

I bring to this text not simply a close reading, but political commitments and personal attachments. As I mentioned earlier, I became involved with the *Maya Atlas* in 1995 when I assisted in writing the grant proposal for the "Maya mapping project." I was later invited by the TMCC to collaborate with the geographers from UC-Berkeley to train the community researchers and produce the *Atlas*. I was involved in many of the discussions about the planning of the project and participated in all of the workshops.[16] Therefore, this chapter presents a critique of a text that is partly my own. I recognize that concluding my book in this way may reflect narcissism, but I think the importance of the *Atlas* – read against the archaeology of Mayanism – is evident (and, in any event, I was only one of many *Atlas* authors). Still, any argument for reading the *Maya Atlas* would be shaped by my nostalgia for the time when the *Atlas* was made, when I participated in producing a vision of collective landownership that has yet to be fulfilled. The *Atlas* is both a text and an incomplete political project, as is suggested by the text's subtitle: *The struggle to preserve Maya land in southern Belize*. I remain involved and implicated in this struggle, and its maps.[17] Postcolonialism is one name for the effort to carry forward such political struggles without lapsing into nostalgia.

Making the *Maya Atlas*

> This is the first time anywhere that indigenous people have made their own atlas.
>
> B. Q. Nietschmann (1996)[18]

The concept of mapping the Maya lands of southern Belize predated the making of the *Atlas* by at least a decade. A sketch map of the "Maya homeland," drawn by the project's two main leaders, Julian Cho and Diego Bol, had been drawn in the mid-1980s.[19] These same two leaders were central to the *Atlas* project: Bol was the *Atlas* project coordinator, Cho the political leader that made the project possible. As we have seen, in 1995 Julian Cho was elected Chairman of a new group commissioned to investigate the land rights situation (the Reservation Lands Committee: see chapter 5).[20] Each Sunday, the six members of the committee would speak in two communities on the status of the Indian reservations and argue for a movement for collective, indigenous land rights. These Sunday meetings were the crucible for a number of momentous changes in the Maya movement. They provided regular opportunities for public analysis of the land issue; it allowed the movement leadership to build support against the logging concessions in the CRFR; and it led to the identification of Julian Cho and his committee as the core leaders of the Maya movement. When the TMCC held their biannual assembly on December 31, 1995, Cho reported on the committee's findings and announced plans to map the Maya lands. When the time came to elect a new TMCC Executive Council, the committee members swept to power, and Cho was elected Chairman.[21]

With Cho as Chairman, the TMCC deepened its institutional capacity and initiated a number of ambitious projects. An office was established in Punta Gorda; the campaign against the "Malaysian" logging company grew; negotiations with the government were initiated; and grant proposals were sent to international funding sources. From these efforts emerged the Maya mapping project, which evolved into the *Atlas*, and a lawsuit against the Government of Belize to win land rights under the provisions of international indigenous rights law.[22] Cho developed a close working relationship with Santiago Coh, the Chairman of the TAA, leading to close collaboration between the TAA and TMCC on both the lawsuit (of which Coh and Cho were co-signatories) and the co-authorship of the *Maya Atlas*. Cho and Coh traveled together to Washington, DC, in April 1996 to meet with the Indian Law Resource Center (ILRC) and Nietschmann, whose team they selected to manage the technical aspects of the mapping project. Upon their return, Cho and Coh held a press conference at Belize's international airport, where Cho announced their plans:

> The number one plan that we have is to begin a mapping process of all Maya inhabited areas in the Toledo District. This would be to find out where the Mayas live. We have done this with consultation of the people and [Chairman of the Alcaldes Association] Mr. Coh can tell you that we have contacted the thirty-four villages. We are going to begin the mapping process and it will take some time. We are looking forward to getting cooperation from the government, but even without the government cooperating with us...we will be able to get some technical assistance abroad.... I want to tell the government that we can go abroad, and we can lobby, and we have a lot of friends abroad, and that if they don't listen to us, then it will get rough in Toledo.[23]

The TMCC and TAA fulfilled Cho's promise to map "all Maya inhabited areas in the Toledo District" within four months (see figure 6.1, one of the community maps). The maps, photos, and text were then taken to Berkeley to be edited, and the original hand-drawn maps were photographed, scanned, and combined to produce a map of "Maya land use in the Toledo and Stann Creek Districts" (see figure 6.2). Draft copies of the *Atlas* were carried from Berkeley to Punta Gorda in June, 1997, where final layout decisions were made. Nietschmann worked with the *Atlas* publisher, North Atlantic Books,[24] to prepare the final text in Berkeley; the *Atlas* was printed in Hong Kong and released in 1997 in Belmopan.

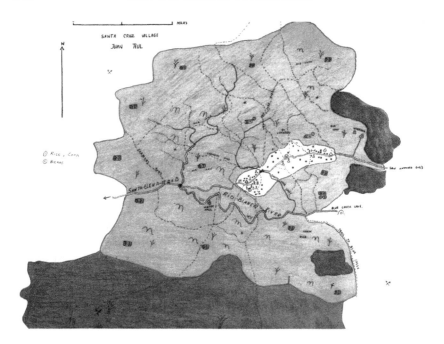

Figure 6.1 Map from the *Maya Atlas*, 1997

Source: Maya Atlas: The Struggle to Preserve Maya Land in Southern Belize by Toledo Maya Cultural Council and Toledo Alcades Association. Published by North Atlantic Books © 1997 by Toledo Maya Cultural Council and Toledo Alcades Association. Reprinted by permission of the publisher. The village researcher for this particular community is Juan Teul.

One can find many criticisms of the *Atlas* project in southern Belize.[25] Foremost of these concerns was the difficulty of finding a copy of the *Atlas* in the communities it represents; most Maya people in those communities have never had an opportunity to read the *Atlas*. The *Atlas* costs Bz $60 (US $30),[26] but copies were supposed to be available for free in the Maya communities. Although the TMCC distributed one copy to each community in 1997, nearly all of those copies have been lost or destroyed. Some who worked on the project have expressed to me that, given the cost and difficulty of the project, the local effects on TMCC's work were negligible. This is itself a powerful criticism of the *Atlas*, insofar as the text claims to represent the Maya: the *writing* of the text was more participatory than its reading. Many Maya people first encountered the *Atlas* when the Environment, Social, and Technical Assistance Project (or ESTAP) conducted community-based workshops in 1998–9.

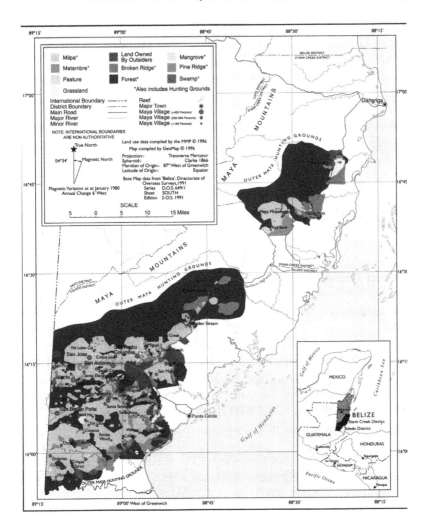

Figure 6.2 Map of the Maya Cultural Land Use Area from the *Maya Atlas*, 1997

Source: *Maya Atlas: The Struggle to Preserve Maya Land in Southern Belize* by Toledo Maya Cultural Council and Toledo Alcades Association. Published by North Atlantic Books © 1997 by Toledo Maya Cultural Council and Toledo Alcades Association. Reprinted by permission of the publisher.

Ironically, ESTAP was created because of opposition to the paving of the southern highway by the TMCC. Thus, after all of the challenges that the TAA and TMCC overcame to make the *Atlas*, its first systematic use in the communities came through a development project they initially opposed.[27]

Two other criticisms are prevalent in southern Belize. The first is that the *Atlas* contains factual errors, especially concerning the origins of particular communities. Second, the *Atlas* and its maps failed to shift the political terrain in the struggle for Maya land rights. These are valid criticisms that no one involved with the project should deny, but they only go so far.[28] A postcolonial reading is still necessary to consider the possibilities for broadening the critique of cartography and Mayanism.

Reading the *Maya Atlas*

Let us turn to the text, beginning from the section on land use.[29] The section begins with the map of Maya land use and a glossary of terms related to Maya land use. *Milpa*, we are told, is a "plantation of corn or rice done by the slash-and-burn method."[30] The section on Maya land use begins by discussing agriculture and posits several claims: that "Mayas live in a communal land system" that is "still run by the rules of the traditional Mayas";[31] that these rules are not recognized by the state; and that leasing is unknown:

> Mayas live in a communal land system. The Maya system is basically still run by the rules of the traditional Mayas, although this is not recognized by the present Belizean government. The Maya do not know about lease land. This information is for today's children of the Maya.

Why do *these* statements introduce the section on land use? What is at stake here? The first thing to notice is that all of these claims are questionable. For instance, while it may be true that there are areas that are communally held in southern Belize, there are few such spaces remaining. This is one of the main reasons for making the *Atlas*: to produce some kind of a "communal land system" that is managed along "traditional Maya" lines. Further, the basic practices that constitute the "Maya farm system" have been carefully studied for about a century by the state. Whatever else we can say about the state's interpretation of Maya landholding and *milpa* practices, we cannot say that the "present government" or any other has not recognized its existence.[32] And land leasing has a long history in this area. Indeed, several of the Maya leaders involved with the *Atlas* were leasing land at the time the text was written.

How, then, are we to interpret these statements? Are they simply false? Are they errors? Are they true, but only within the Maya cultural framework? I propose that we read these statements as reflecting a set of desires. For instance, we can read the statement "the Maya do not know about lease land" as a reflection that the authors do not *want* to lease land, but do. Why not: "The Maya often lease land, but we hate it; we would rather have a Homeland." Similarly, why write "Mayas live in a communal land system," and not, "Mayas no longer have a communal land system; we would like one"? This would seem to be closer to the truth as well as the stated aim of the *Atlas*. The paragraph's last line provides an important clue for our reading: "This information is for today's children of the Maya." The three claims that preceded this statement were therefore intended to be for the children of today's Maya, that is, the authors' descendants. The most curious word in the paragraph is "information," which suggests that we are talking about data, or facts, that should be given to the children of the living Maya when they are older. But these are not mere information.[33] These are statements about the essence of the Maya land system, which, if we take the *Atlas* as a whole, means that they concern the basic character of what it is to be Maya.

Let us continue to read the section on land use. The *milpa* system is briefly described; land management is said to be governed by the alcalde; the process for land selection and demarcation is summarized; and then conflicts over land are discussed:

> We have caves, hills, and rivers to be used as demarcation lines that separate one community from another.[34] If a problem arises, first we consult the leader of each community. We argue among ourselves about the demarcation lines.... For every problem that affects all of us, we consult each community leader. A joint solution is found. Then we educate the villagers about the discussion. When non-Mayas have a problem with us over land, we have to solve it very differently. For example, a foreign citrus company began operating in the area of San Felipe village. We were against this, because it is our land. But when the Government Lands Office was consulted concerning the matter we were told that the land was not ours, that it is not even a reserve, but crown land.[35]

This pointed tale offers a view of how state-society relations are imagined. Land problems "internal" to Maya society are worked out through argument, consultation ("consult" appears here three times), and education; land problems caused by "foreigners" lead to engagements with the state, and are rarely solved. The approach to land is bifurcated along the lines we/foreign. The question this raises is how these divisions, and the "we" of this statement, were historically constituted. It is a collective "we" that is clearly produced as an imagined community.

The remainder of this brief section entails a description of corn planting, including a lengthy discussion of the ritual prayers and practices central to the corn planting.[36] It begins: "The Mayas plant corn the way our elders taught us." After saying the names of the sacred hills and mountains, "The man begins to plant, taking a deep breath because he is sorry about piercing the face of the earth with a planting rod."[37] The section focuses entirely on corn. The only discussion of any other crop comes in a short supplementary section, entitled "More on Maya agriculture," that simply names a few other plants. The section ends with a statement that captures the essence of the land use section: "In these modern times the Maya people are expanding agriculture by planting citrus or cacao orchards and raising cattle, but not forgetting their traditional way of life."[38] Thus the whole question of land use is cast in terms of a struggle to remember and maintain traditions in the face of modernity.

Before going further we must note that the 1,000 or so words are not the only thing that comprise this section on agriculture. The text is framed by six color photos and a drawing. The latter is said to portray "Yum Kax – corn god."[39] The photos show two cornfields, a stack of corn, a farmer with a horse, a stack of rice stored inside a house, and a path to a plantation. With the exception of the photo of the stored rice, all of these images, like the text, emphasize corn production. There is no mention of rice production, although rice has been the most important cash crop in southern Belize since at least the 1970s. Why not mention a crop so fundamental to livelihood? The photos and the text together suggest that corn is the essence of Maya land use, identity, and survival. And they suggest its means of production: a man walks through a path to the *milpa*, says his prayers (presumably to Yum Kax), works, carries the corn home on his horse, and stacks it in the manner represented in the photo (the caption reads: "traditional corn storage").[40]

Given the archaeology of the Maya farm system discourse, not to mention the teleology of the tradition/modern structure, we cannot accept this representation of land use and agricultural practices as an innocent, objective statement of the facts. Nor can we simply read it as an honest reflection of a culture that happens to be determined *by* that culture, since we are, after all, reading in English after more than a century of British colonialism and Mayanist discourses. In other words, we cannot presume that the reason rice is not mentioned in the section of land use has to do with some essential "cultural" reason for not discussing it. Our reading must pursue the production of "Maya culture," not take it as a starting point. In so doing, we see that "Maya culture" is marked through certain signifiers: corn; green hills; machete; Yum Kax; horse; plantation path.

Each of these terms works – produces the meaning of "Maya culture" – by marking off other possible signifiers. For these, important exclusions include: rice; urban spaces; chainsaw; the city; the Christian God; pickup truck; road. These are certainly not the only possible signifiers that could be paired. My point is that rice, chainsaw, and God could be used as equally "authentic" and "truthful" (if nonetheless ambivalent) signifiers for present day "Maya culture." The question is not therefore one of mere sloppiness or ambivalence. Nor should we assume that a traditional Maya culture is losing its essence to "modernity." Rather, the meaning of "Maya land use" produced in the section on land use is constituted by the play of the presence and absence of signifiers for which there is no preexisting or correct guide. A critical or postcolonial reading will find that certain tropes and signifiers that constitute "the Maya" are emphasized, while others are deemphasized, erased, or framed in a negative light. It is the play of these signifiers we should track, since it is in their movement that what is "Maya" is constituted. If an exhaustive, rigorous reading were possible, one could move through the text to discern how the Maya subject is constituted through the play of these oppositions – none of which are ever settled.

I have attempted to track these oppositions. To present my findings – albeit in a didactic form – I offer the following pair of lists (see figure 6.3). The left column lists signs that are emphasized in the *Atlas*; the right column suggests what must be excluded for the left column to work. I call these "antinomies" because they are pairs of "opposites" that do not necessarily "oppose" one another, except in the sense of producing meaning. Neither column can be read as a reflection of what is genuinely "Maya," since all of these *could* be used. No one could deny that Maya cultural practices in Belize today concern both machetes *and* chainsaws, horses *and* trucks, rural communities *and* urban spaces, etc.[41]

To clarify what this table shows us, let us briefly examine two themes from this list. The first relates to a set of pairs involving *livelihood*:

farming and hunting   "jobbing out," semi-proletarianization
collective labor, moral economy   wages, competition

When the *Atlas* discusses Maya livelihoods, it stresses that Maya people share labor with one another, through the *fajina* system and the use of collective labor teams. (There is a considerable ethnographic literature on these labor sharing practices among the Maya.) What goes unmentioned in the *Atlas* is that such practices are neither universal nor evenly

horses, saddles	pick up trucks
machetes, axes	chainsaws
the marimba, the harp	electric guitars, raggae music
corn and cacao (subsistence crops)	rice and marijuana (cash crops)
sustainability, abundance	poverty, ecological complexity
hand-made crafts	commodities, commercial exchange
those who came from Guatemala	those who were already 'in Belize'
distance to Punta Gorda	distance to Coban or San Luis Peten
view from above: horizontality	view from the ground: verticality
the 'community' scale	other scales (household, region, etc.)
extent of land use	intensity of land use
roads and rivers	air routes, telephone networks
smooth landscape patterns	uneven or complex landscape patterns
rural village life	urban life ('Indianville', Belize City)
Homeland, communal lands	privately-owned land, leased land
ancient Gods	*Dios*, Christian God
ancient Mayas: parallels	discontinuity with the ancient Maya
Belizean Mayas	Mayas in Guatemala and Mexico
Q'eqchi' and Mopan Maya	Yucatec Maya, Creole, Garifuna
the Maya flag	the Belizean flag
legal strategy	mass politics
logging conflicts	religious conflicts, village politics
'The Government'	party politics, the PUP and UDP
'bush medicine'	the hospital in Punta Gorda, health care
farming and hunting	'jobbing out', semi-proletarianization
English language text	Mopan and Q'eqchi' texts
making the Atlas in Belize and Berkeley	printing the *Atlas* in Hong Kong
collective labor, moral economy	wages, competition
Founding fathers	followers, deportees, predecessors
Maya traditions	colonialism, historical struggles
continuity, stability	change, discontinuity
communities, rootedness, settled life	migrations, movement, shifting, *alkilos*

Figure 6.3 Antinomies of Maya culture in the *Maya Atlas*

employed today. Generally speaking, non-wage labor sharing practices are most common with subsistence, noncommodity production (especially corn). If a farmer is planting a cash crop, such as rice, it is not uncommon to hire wage labor. And women's groups that produce crafts together will expect to be paid in cash if some of their crafts are taken to be sold by a member of their group. Moreover, wage labor has been a substantial livelihood strategy for Maya men in southern Belize for at least a century. This fact was not only well known by the representatives who made the *Atlas*; it was well documented in their research. The *Atlas* community research survey found that one out of four adults were regularly involved in *milpa* farming, and a few more than this were involved in raising animals (because women and men are often both involved). But it found that one of out five men had tried "jobbing out," or finding wage labor outside of their home community. In other words, the research suggests that *milpa* farming is only one livelihood strategy (albeit a crucial one for subsistence reproduction) in a complex and dynamic ensemble of strategies in the communities.[42] But it is the only one that is really addressed in the *Atlas*. If finding profitable and meaningful employment is a central challenge in many people's lives, then why not address this?

Second, consider the treatment of the historical geography of the area. One of the fundamental reasons for conducting the research behind the *Atlas* was to document the popular understandings of the region's social history. The community survey included questions that were intended to draw out the memories of settlement and historic patterns of land use. This material would help to ground the legal claim to longstanding cultural rootedness in the area. In this respect, the research project largely failed. Very little new material was collected. Although the *Atlas* suggests that the living Mayas trace their ancestry back to pre-colonial Maya settlements, the village histories that accompany the maps stress the recent dates of settlement, often naming the "founding fathers" responsible for founding villages in the twentieth century. What is left undisclosed is the tragedy of the "three conquests" of the Maya – the experiences of war and forced settlement during Spanish colonialism, the nineteenth-century liberal reforms, and the civil war in Guatemala. British colonization is similarly neglected. Given the absence of statements by rural Maya peoples about "their history," the story that is told is a largely happy one of settlement in stable communities:

Maya traditions	colonialism, historical struggles
continuity, stability	change, discontinuity
communities, rootedness, settled life	migrations, movement, *alkilos*

There is a strong spatial character to this pattern. The Maya community centers and roots. It does not splay and shift. It is remarkably settled.

Consider the discussion of *alkilos*, mentioned in several village histories. The opposition of settled communities (which are privileged in the *Atlas*) versus *alkilos* was a key trope in British colonial discourse about the Maya.[43] Compare the three definitions of *alkilo* in the *Atlas*: Santa Cruz "was first an *alkilo*, meaning that people lived in the forests far from each other in no particular order." In Santa Elena, it is written that the "founding fathers first called this settlement Rio Blanco after the crystal-clear river that bordered their *alkilo* (scattered Maya houses in the forest)." Na Luum Caj, we read disparagingly, "still resembles an *alkilo* (a style of Maya community in which the houses are situated at a distance from each other)." In each case, the existence of *alkilos* is accepted, but always in a negative light, as an undeveloped moment in an evolutionary sequence. Santa Cruz and Santa Elena may have been *alkilos*, but they have graduated to become settled communities. In the description of Na Luum Caj, the keyword is "still": that "houses are situated at a distance from each other" is admitted as a kind of problem. Thus *alkilos* exist as a negative reference to the more permanent "communities." The *Atlas* never reveals the location of existing *alkilos*, and the *alkilos* and communities that are not presently inhabited (such as Xpicilha) are never mentioned. If logocentrism names the prioritization of the Logos in the history of Western metaphysics (as Derrida argues in *Of grammatology*), does the treatment of the *alkilo* in the *Atlas* not suggest that such logocentrism collaborates with a spacing that centers, that prioritizes that which is ordered, sorted, and centered? Is there a logo*centrism* at work in anticolonial discourse? If so, how can we resist the linkage of Reason, science, and settled presence? Why not speak to the strategic ways that the Maya people have overcome colonialism through fluid resistances, movements, and dispersions?

What do these cases, and figure 6.3, demonstrate? First, although the singularity of this text would seem to suggest otherwise, there could have been many different *Atlases*. We could imagine an *Atlas* that emphasized the heroic struggle of the Maya people against the colonial invaders. But this is emphatically not the narrative of this *Atlas* (which makes almost no mention of colonialism) and it would be no less anti-colonial. Second, the *Atlas* emphasizes a kind of imagined, homogenous "Maya tradition," one constituted through practices that the men involved in this project saw as essential to Mayaness. Third, this smooth and consistent community essence is spatialized through the text. What constitutes Mayaness is not read through the unceasing flow of bodies, practices, and colonial encounters over the past five hundred years. Rather, it is in the settlement of stable, rural, agrarian, bucolic, and self-identified "Maya villages" that the essence of Maya culture resides. In other words, through the play of the presence and absence of certain crucial signifiers, the *Atlas* provides a way of constituting Mayaness as an essentially spatial complex of cultural-national phenomena. The meaning of what constitutes "Maya" space in the *Atlas* is produced through a set of exclusions, the kind which are necessary for territorial, national identities to be constituted as such.

Space and Gender in the *Maya Atlas*

The clearest evasions and points of intensity in the *Atlas* concern the representation of sex and gender. The *Atlas* claims to show us the Maya world, but it is a world of Maya men. This should come as no surprise, since the group that created the *Atlas* was made up exclusively of men.[44] They regarded themselves as representatives of the broader "Maya community" and produced the *Atlas* to "reflect" the Maya world. But this clearly causes anxiety in the text, since women are absolutely necessary in the representation of the Maya world. The resulting treatment of sex and gender difference produces a sense of what it means to be "Maya" through the abjection of women's spaces and lives. Women are not simply excluded, but are rather included in such a way as to be marked off, separated, and silenced through representational practices that are neither strictly "negative" nor erasures (the absence of women altogether).

Consider again the maps of the "Maya communities" (e.g., figure 6.1). These maps present southern Belize not simply as Maya, but also male. Spaces where men are productive – corn fields and hunting grounds – are equated with Maya livelihood and territory. They are colored various shades of green. By contrast, spaces where women work and live – including the household, riversides, and village paths – are represented as small black dots, blue squiggles, and black dotted lines. Thus, the *scale* of the community maps privileges men's practices as farmers, wood-gatherers, and hunters. While the reader can easily make out farmlands, hunting grounds, and wood stocks, it is very difficult to see where women do their work. The spaces where women live and work are drawn in a way that leaves them not only small, but unexamined. By delimiting the scale and through the uneven use of color and detail, the maps represent "Maya community" in a way that celebrates men's movements and work while treating women's spaces as static and aterritorial.

We should ask whether this representation of women's spaces does not presuppose a prior assumption about the "proper" gender roles in Maya society. It does – and this presumption of what constitutes appropriate gender roles and spaces is contested. But the politics around the appropriate spaces and roles for men and women are not discussed in the *Atlas*. Rather, the *Atlas* presents a view of Maya society as a bifurcated space with a balance of responsibilities for men and women. We should call this oversimplified view of Maya social life into question; but first we should note the way that this division is not simply split but weighted, with all that is "male" figured as normal, and "female" as abnormal, different, and traditional. For instance, early sections of the *Atlas* discuss the organization behind the *Atlas* (with men working through the TMCC and TAA), the history of the region (men founding villages), and land use (men planting corn). The so-called women's issues appear in the *Atlas* under the heading "special topics."[45] When the black dot that represents the household is entered, through the photos of women labeled "special," they only portray women fulfilling what are perceived as traditional Maya customs. Above all, women are shown making crafts. I note that these are tasks that are taken to be traditional because, of course, what the text is doing by showing Maya women at work is constituting "Maya tradition." (For instance, women are shown weaving jippi-jappa baskets, a commodity sold to tourists of recent production in southern Belize.)[46] So the problem is not that women are simply excluded from the *Atlas*. This would be impossible. The *Atlas* is so tense, so anxious, to *include* a discussion of women that they are given their own, "special" section. Women are thereby included in a way that renders them voiceless: a necessary point of departure; an abject marker of a "Maya tradition" that they embody but cannot speak of.

The relevant question is how the *Atlas* contends with women, sex, and gender. Norms of gender relations enter the text in several ways: through descriptions of marriage rites; photos of heterosexual couples; and through heteronormative statements about Maya families. The *Atlas* does not give us a sense of what life is like for the unmarried, widows, or queers. Nor is there any mention of spousal abuse, arguably the most prevalent mode of gender-based violence in southern Belize.[47] The abjection of women and girls that makes such violence acceptable is reiterated through the *Atlas*, especially in the photographs. This is particularly visceral in the case of a photo by Nietschmann: a young women is shown, standing with her back to a wall. Her demure smile is framed by hand-woven belts – a kind of "Maya handicraft" made for sale to tourists – which hang along the wall. Some of the belts have been arranged to fall in front of her, as if she is tied by the belts. The twisting belts suggest that she is restrained, sexually available, and simultaneously other, "exotic." In this image we find the collision of all that is required of Maya women for the *Atlas*: an affiliation with tradition, crafts, and sexuality.[48] The image of woman is bound to the home, a small black dot, by crafts of her own making.

If we recall that the overarching reason for making the *Atlas* was to win land rights, we must ask: what is at stake in the scaling of "Maya land" in terms of the men's movements and production? Is there any doubt that men would define and control the "Maya homeland"?

Counter-Mapping as Territorialization

> The Maya Mapping Project and the *Maya Atlas* were designed to be an assessment of the natural and human resources of the Mayas' proposed Homeland. To govern a Homeland it is necessary to know what is there to govern. Therefore, the Maya Mapping Project would map and inventory the Homeland communities and community lands.
>
> TMCC and TAA (1997: 138)

What happens after all of the "Maya lands" are mapped? If counter-mapping projects from Indonesia and Tanzania are any guide, once lands are mapped, surveyed, and bounded, access will be restricted in new ways. Community managed "national land" becomes, if not a fictitious commodity, at least more likely to become property. This is why the complex politics of the worlding of "the Maya world" is so intensely political: the territorialization of land through counter-mapping is sure to displace or map land claim disputes onto other conflicts. Because participation in the process of producing maps and control over the new forms of spatial information are certain to be uneven, there is no way to ensure that counter-maps could fully represent subaltern geographical knowledges. Since counter-mapping is usually done in the name of a particular cultural or national community – as in the case of "the Maya" for the *Atlas* – the way claims are spatialized will reflect the broader aims of those "representing" that national identity. Representation is always of a twofold character (as re-production and speaking for), as Spivak shows in her reading of Marx. What is at stake in counter-mapping is the spatiality of national representation in both senses: the "speaking for" and reproduction of a *spatialized nation* are articulated through the ontological formation of territory.[49]

Territorialization always involves the production of space in such a way that it has the character of belonging to a nation-state. This requires worlding space in a way that insures an orientation, directionality, scale, and so forth that are congruent with the recognition of a national quality and state governability. All of this presupposes entering into space as a calculable region defined by coordinates in the "modern world." In other words, it requires a worlding and coordination of the type characterized by Descartes (in what Heidegger argues reflects the manifestation of the whole history of the Western metaphysics of being as presence since the Greeks).[50] Such spatiality is not only subordinate to time in Western metaphysics; it is ultimately measured by an absolute time that Heidegger calls "vulgar time."[51] Things are located in terms of their distance, in "minutes" and "seconds," relative to Greenwich, England. Such a spatiality not only presupposes a positivist empiricism (all spaces can be exactly defined through measurement); it is also solicited by positivism, which desires to be definitely located.

To map the Maya world, the Maya mapping project had to start some*where*. One maps a world, but never anew, and never with or within a space made *de novo*. There always already *is* world.[52] One maps with orientations, projections, and units at hand. These are the grammar and syntax of the writing of the world. They are more than bits of worldliness that one has "in mind." They are things with materiality. Even if there are no maps or spatial data at hand when one begins to map, no maps at hand when "the pen touches the paper," one cannot work free of worldliness, orientation, scale, units, or projection. And especially when the mapping is a project of producing *origin* (as in the case of the *Atlas*) the question of *orientation* is of special importance. The desire of every cartographer is not to write a prethematized world: it is to rethematize the "real world" for certain purposes. Hence the unusual textual practice in cartography of producing truth through the marginal annotation of error.[53]

Could the Maya mapping project get outside of the Western dream of a firm, uncontested, positivist spatiality, tied as it is at the root to exactness, geography, calculatability, property relations, capitalist development, the metaphysics of presence, territory, and so forth? Are all of these already written into the map? To say this differently: can the subaltern map?

Before answering this crucial question, let me briefly summarize how, in fact, the counter-maps of the *Atlas* were produced. I have explained that the community researchers, once selected in their communities, were trained in basic cartographic skills. And they were trained to see their communities from the aerial view preferred by Western realist cartographers. They were taught how to place the things that define their communities (green fields, black dot houses, etc.) onto a blank sheet of acetate, so these symbols could be transferred later to a GIS. But how would all these different community researchers orient their fields and black dots in exactly the same way? The fundamental cartographic problem: what precise scale and projection should be used to provide the essential framework for these spatial representations?

The solution to this problem – the *only imaginable solution*, in fact – was to begin with *another map* that was produced on the basis of an exact and acceptable scale, projection, and so forth, and to copy enough material from it to make a common set of semi-reproductions that could then be elaborated. And so, as the starting point for the territorialization of Maya land, the community researchers traced rivers and roads from another set of maps, called "base maps." Thus we find, in a marginal note on the list of maps in the *Atlas*, the following "metadata":

All data compiled by Maya Mapping Project (MMP) © 1996, Basemap Data from (1) *Belize* 1:250[,]000, Directorate of Overseas Surveys, 1991, series D.O.S. 649/1, sheets North and South, edition 2–O.D. 1991 or (2) *Belize* 1:50,000, Directorate of Overseas Surveys, 1991 series (DOS4499), Sheets 30–44.[54]

In other words, the maps in the *Atlas* are literately drawn on those maps, entitled "Belize," made by the British Directorate of Overseas Surveys in 1991. The spaces of the Maya world are already marked by the trace of the Empire: the nation-state-territory known as "Belize," as drawn by British surveyors. Written on the margins of *these* DOS base maps – but not, notably, reinscribed on the "Maya" maps drawn on their basis – is the following:

Grid: UTM Zone 16
Projection: Transverse Mercator
Spheroid: Clarke 1866

Units:	Meter
Meridian of Origin:	87 west of Greenwich
Latitude of Origin:	Equator
Scale Factor at Origin:	.9996
Datum:	1927 North American

I noted earlier that maps are distinguished from other texts in that they display marks that qualify their truthfulness on their margins. Maps produce truth by accommodating their inaccuracies and distortions (their "untruth") in a rigorous form of marginalia of metadata like these. Mapmakers know that there can be no perfect representation of the world; indeed, cartography could be defined as a form of techno-poetic writing that produces imperfect spatial representations of a world – our world that resists two-dimensional, reductive reproduction. The marginal notes about the grid, projection, spheroid, units, scale, and datum therefore play a double role. On one hand, they are clues to the reading of the map: the mathematical rules to the flattening of the world, guides to measuring the relative distortions of distances, shapes, and angles. In this way, they are the openness that promise the trustworthiness of the text. Insofar as the rules for distortion are given together with the map image, they suggest the unrepressed, even self-reflexive presentation of the conditions of production of the text. This is the objective presencing of the practice of distortion. By providing the metadata for all to read, the cartographer offers up the inevitable error for consideration.

In this way, these metadata are an exemplary metaphor for the problem of the Western notion of objective truth: by writing the limits of knowledge into the text, the cartographer believes to have resolved the problem of a subjective residue in the worlding of the world. But this objective marginalia only reinscribes a more fundamental objective view from nowhere. This is belied by the spatial lives of these metadata: these maps are enframed by a grid drawn onto the Universal Transverse Mercator projection; the units are in meters; the meridian of origin is drawn in relation to Greenwich, England; the datum is based on the 1927 North American calculations; etc. There is already worldliness here, and not only in the "objective" presencing of an abstract positivist, exact spatiality.

Given the form of power-knowledge embedded in *these* maps, which were literally made by the colonizing army, why use them? Are there no other starting points, no more innocent points from which to see? Are there no *other* maps from which to begin? Yes, there are always other maps. The question, or better the aporia, is how *other* they are.[55] This aporia is not only limited to questions of scale, projection, and symbol. All of these are, in a sense, extensions of the condition that spatiality is the *différance* of cartography. As a mode of writing, cartography must work on the basis of an inevitable differentiation-deferring of all that is not posited, oriented, worlded, and circumscribed. In the case of the "modern" world, such circumscriptions inevitably involve boundaries that correlate, however violently, with the nation-state-territory trinity.

The counter-maps of the *Atlas* are no exception. During the making of the *Atlas*, the project's leaders debated how to represent the boundary line between the countries of Belize and Guatemala on the maps of Maya villages which lie near the border. Some leaders were concerned that emphasizing cross-border movements and exchanges (which are very common) would diminish the power of their claims to political rights as *Belizean* Mayas. Therefore the decision was made to make no mention of the ways Maya people in Belize frequently enter Guatemalan territory, and to make no mention of transnational social relations between Mayas. Consequently, the community maps from those villages that lie near the border reflect a D-shape that belies the cross-border flows of everyday social life (see figure 6.4). The boundaries around the maps thus reinscribe a spatial Belizeanness even as they try to define a space that is aboriginal to Belize. The desire to qualify as national citizens while staking indigenous claims to territory is an aporia that characterizes the postcolonial situation. What is at stake in the positing of a space as "aboriginal" land is the becoming-space of those nations that are always already "in" colonial space (territory) and temporality (modernity). The category "aboriginal" constitutes the becoming-space of colonialism, of the colonial nation's history. Postcolonialism is one name for a reading that tracks the production of this subject and the conditions of this becoming-space of colonialism.[56]

To back-step momentarily: we have seen that counter-mapping must begin somewhere, and that in this case, in defining this "where," the anticolonial cartographers of the *Atlas* began with Cartesian space and colonial air photos. How we read this problem is important. We could treat the starting with the British Royal Air Force (RAF) maps as evidence that such colonial maps are "ambivalent at the origin," in Bhabha's sense – we could say that the spacing that makes anticolonial knowledge possible is produced with texts from the colonial archive that are ambivalent at the origin. This might lead one to hope that there are already spaces in the maps for a different anticolonial politics, or that in every colonial map there is an anticolonial map crying to come out.

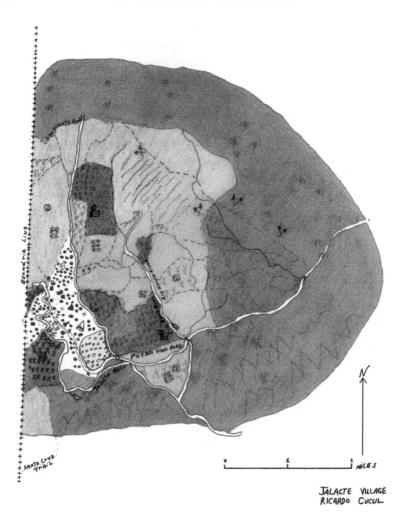

Figure 6.4   Map from the *Maya Atlas*, 1997

Source: *Maya Atlas: The Struggle to Preserve Maya Land in Southern Belize* by Toledo Maya Cultural Council and Toledo Alcades Association. Published by North Atlantic Books © 1997 by Toledo Maya Cultural Council and Toledo Alcades Association. Reprinted by permission of the publisher. The village researcher for this particular community is Ricardo Cucul. Note how the village center, shown in white with houses marked as black dots, appears to open itself to the Western border with Guatemala. The community's socio-spatial life in Guatemala is not addressed or mapped in the *Atlas*. This is one of five such "D-shaped" maps in the *Atlas*.

I think we should reject this reading, for reasons suggested by Heidegger and clarified by Derrida. As I have argued, the apparent valences of a modern map (rarely reducible to two anyway) are themselves made possible by a prior spacing, one that cannot be read apart from the history of Western metaphysics. This problem gives evidence to Heidegger's claim that in the worlding of the world, the becoming-space of vulgar time is Cartesian. The RAF photos and the colonial maps constitute the necessary "points of departure," literally, points on maps that are nothing but the differentiation-differing of space that make possible certain geographies. Such a beginning, or starting-orientation, is necessary; it is a condition of all maps, counter-maps included. We must see this even as we recognize that this starting-orientation undermines the possibility that the counter-map ever will fully present or counter the worldliness of the world in the way that we desire it to. (This is why there can be no postcolonial critique of social history without a simultaneous deconstruction of the ways geography and cartography are already implicated in colonial knowledge. Without a critical reading of the becoming-spaces of history, we are led to hope, in Heideggerian fashion, that the problem of ontology is to be worked through only on the horizon of temporality.)

Counter-Mapping as Cultural Ecology

If you're interested in cultural diversity, you have to be interested in biological diversity, because nature is the scaffolding of culture – it's why people are the way they are.

B. Q. Nietschmann, quoted in Sanders (2000)

I will return to this theoretical juncture, but here should return to the *Atlas* and the challenge of reading it as a text of postcolonial geography. I do so by returning to the question of Maya identities. A few months before the *Atlas* was to go to the publishers, Nietschmann wrote to the leaders of the TMCC and TAA to ask them to generate more "text":

The *Atlas* does not have much text. We need a lot more writing by the Maya. I suggest it should be done as if you were writing to explain Maya life to the next generation of Maya children. It shouldn't be speeches or writings for the press. This is because the *Atlas* will still exist after the land question is resolved. It will be used to help govern and to help educate Maya people about all the Maya communities. So we need more on Maya history in the region, the settlement history of the communities, the transcriptions from the tape recordings, and stories about corn, traditional healers, women, Maya food recipes, a description of what village people do during a year, month by month. These things.[57]

The first thing to note about this extraordinary statement is that Nietschmann does not simply say that more text is needed. It is not that the TMCC and TAA are left to decide how to fill the blank spaces of the *Atlas* with anything that they wish to write, draw, or say. Rather, Nietschmann determines the task – "to explain Maya life" – and goes on to define what this includes. Not surprisingly, the list reads like a page from the methodology section of a cultural ecological dissertation: "stories about corn, traditional healers, women, Maya food recipes, a description of what village people do during a year." Tradition, food, women, and circular time: if these are the keystones of Maya life, why do the Maya leaders need Nietschmann to specify them?

To answer this question, we need to consider the relationship between counter-mapping and Nietschmann's approach to geography and cultural ecology. It is no accident that Nietschmann, a major proponent of counter-mapping and technical leader of the *Maya Atlas*, was also a legendary cultural ecologist. "Counter-mapping" is grounded conceptually by cultural ecology. The essence of cultural ecology is captured by Nietschmann's statements on the relation between culture and the environment: cultural ecology seeks to explain how it is that a culture is shaped by an environment, and how cultural patterns shape the landscape in turn. It is in this sense that, in the long run, nature and society come to mirror each other. Nietschmann's study of indigenous cultural practices and environmental relations, *Between land and water*, is considered a classic work of cultural ecology.[58] The study examines "the ecology of Miskito subsistence and how a particular population had adapted to local ecosystems and modified them."[59] Drawing from intensive field research in one community, Nietschmann develops a model of the cultural-ecological system, albeit one undergoing change as a process of adaptation with capitalist markets. Like others of his generation, Nietschmann's approach to cultural ecology borrowed extensively from the work of ecological anthropologists Julian Steward, Marshall Sahlins, and Roy Rappaport. In an influential essay from 1955, Steward argued that cultural ecology is based on the "determination of how culture is affected by its adaptation to environment."[60] In this view, the task of the cultural ecologist is to empirically define and analyze distinct *culture areas*, spaces constituted through "the utilization of the environment in culturally prescribed ways."[61] Cultural ecology is in this sense a means to produce origin stories of cultural difference. It promises to "explain the *origin* of particular cultural features and patterns which characterize different areas" by rejecting "the fruitless assumption that culture comes from culture."[62] The alternative that cultural ecology proposed was to "develop a *unified theory* of culture and nature, on which could dissolve the culture-nature dualism and replace it with a single totality."[63]

For Nietschmann, it was this complex totality of culture-and-nature that articulated national, natural, cultural spaces as *territory*. The logic is simple and seductive: indigenous traditional land use could be mapped as cultural area; this cultural area indicates the extent of a national claim. We must consider this argument carefully. A presupposition undergirds and connects these two parts of the logic: the equivalence of the *nation, culture*, and *land use*. These assembled concepts are in turn joined as territory. Nietschmann thus felt that cultural practices could be mapped in ways that demonstrate the existence of discrete culture areas which correspond to nations. His theory is clearly explained in *Fourth world war*, where Nietschmann draws a clear parallel between the invading and colonizing forces of Sandinista Nicaragua, Zionist Israel, and apartheid South Africa. The political distinctions between these states were not material to Nietschmann. The influence of this approach to thinking about cultural regions is apparent in the narrative of the *Atlas*. In the *Atlas*, as we have seen, Nietschmann argued that indigenous maps of local land use would reveal, *in a direct and unmediated way*, the areal extent of Maya land use and occupation. This was not a logic that was simply imposed by the premises and requirements of international indigenous rights law.[64] It was inscribed in the very narrative of Nietschmann's "new cartography," in turn grounded by cultural ecology.

In his 1994 essay on nationalism and the "Fourth World," Nietschmann defines a nation as "the geographically bounded territory of a common people as well as to the people themselves."[65] By this definition, a nation is therefore (1) a territorial space, and (2) an encultured people; but it is also (3) the relation between (1) and (2). A nation is the people, the land, and the "people-land." Ergo, nationalism for Nietschmann always implies national-cultural territorialization. He writes:

> A nation . . . is a community of self-identifying people who have a common culture and a historically common territory. And because no group of

people has ever voluntarily given up its territory, resources, or identity, a nation is the world's most enduring, persistent, and resistant organization of people and territory.[66]

Nietschmann's approach presupposes that culture is ontologically prior, real, and rooted in language. It therefore assumes a transcultural and transhistorical concept of nationhood that is itself grounded by specific historical cultural practices: lived languages that mirror environmental realities. However contradictory this may seem, it provides, I suggest, the crucial theoretical backdrop in which to interpret the *Maya Atlas*, since it claims precisely to define the cultural space of the Maya and to elucidate the production of this space as a single cultural-ecological assemblage: "This atlas aims not only to show the boundaries of the Maya Homeland but to bring out the dynamic interactions of the various communities and their relationship to the environment."[67] (This is also why, for instance, Karl Zimmerer locates the *Atlas* within the geographic literature on "ethnoecology."[68] In Zimmerer's words, the knowledge that was produced through the Maya mapping project is "exactly [the] sort of hybrid knowledge" needed for "progress towards ethnolandscape ecology.")[69]

Let us return to Nietschmann's letter to the Maya leaders ("The *Atlas* does not have much text"). What is striking is that Nietschmann *names* the missing ingredients of the *Atlas* – the essence of authentic Maya life, according to cultural ecology.[70] This is why Nietschmann's argument that the *Atlas* "will be used to help govern and to help educate Maya people about all the Maya communities" is troubling. By transferring the duty to catalog the indigenous culture from the anthropologist to the native informant, it works in the spirit of salvage ethnography. Nietschmann scolds these native informants for failing to complete their task: "Many village researchers did not write anything about their community."[71]

An unexpected "solution" to this problem was to allow one of Nietschmann's students, a cartographer, to write the section on the making of the *Atlas*. This option was, of course, opposed on the grounds that it would take away from the fact that all of the text was supposed to be spoken or written by Mayas. Nonetheless, after some confusion about who was responsible for each section, the student wrote much of the final section of the *Atlas*, on its production.[72] This is an ironic metaphor for the aporia of subaltern geographies.

Conclusion

> This impossible "no" to a structure that one critiques yet inhabits intim-
> ately is the deconstructive position, of which postcoloniality is a historical
> case.
>
> G. Spivak (1999: 191)

This chapter opened with a statement by Edward Said to the effect that geography could be made into an art of (anticolonial) war if one could only create "a counter-map." Having read the *Maya Atlas*, we have seen some of the aporias entailed with this task. The only responsible con- clusion to this reading is to recognize, in the spirit of Spivak's interpret- ation of postcoloniality, that counter-mapping is a practice that we cannot say "no" to, and yet must call into question. How can we be certain that anticolonialism will continue to call for counter-mapping? Let us return to Said's call for a counter-map. It was made in December 1993, at a moment when the Palestinian struggle was deteriorating as a result of the so-called Oslo "peace process." In the same essay, Said offers a withering critique of the PLO's leadership and their willingness to sign the Oslo agreements – which abandoned the right of return and instituted de facto apartheid in the West Bank and Gaza – all to win the recognition of a state without any meaningful territory. Said argues that this capitulation resulted from

> the absence of preparation of the Palestinians, who face Israeli experts
> armed with facts, files, and power that have no equivalent on the other
> side. We have been unable so far even to undertake a census of our own
> people. Is there an accurate and usable Palestinian map of the West Bank,
> Gaza, Jerusalem?[73]

The desire for a national map and census extends from anticolonial strategy. Counter-mapping is preparation for negotiation. There can be no anticolonial movement without the maps and land management promised by counter-mapping. The question is whether such a counter- map is always linked to territorialization:

> The struggle over Palestine has been a battle over territorial sovereignty.
> The Zionist idea has always been to coordinate specific concrete steps

with a guiding principle which rarely changes. Thus the Israelis assert sovereignty *and* they build settlements. The Arab technique has always been to make very large general assertions, and then hope that the concrete details will be worked out later. Thus the PLO accepted the Oslo Declaration of Principles on the grounds that Palestinian autonomy would somehow lead to independence if enough rhetorical statements about an independent Palestinian state were made; but when it came to negotiating the details we had neither the plans not the actual details. They had the plans, the territory, the maps, the settlements, the roads: we have the wish for autonomy and Israeli withdrawal, with no details, and no power to change anything very much. Needed: a discipline of detail.[74]

Although Said does not elaborate on what constitutes this discipline of detail, it would be impossible to deny the attractiveness of this formulation. No one familiar with left politics could ignore the importance of such forms of discipline. The problem comes when we recognize that the "details" are already, in some sense, given by that which we aim to oppose. Because what justifies this kind of discipline, in the end, is the need "to survive *as a nation*."[75] And what constitutes the nation in the cultural-ecological mode of counter-mapping is the cartographic overlay of cultural practices – especially language – with territorialized space. The spacing of cultural practices qua nation thereby territorializes by offering the imputed purity of language itself as an ethical basis for drawing lines.

Given my earlier remarks on worldliness and territorialization in the *Atlas*, I think we must accept the responsibility of rethinking the effects of such a model for spacing anticolonial resistance. What if all languages were "essentially" polyglot? What if there were not cultural practices that could cleanly delineate space between nations? Then we would be forced to look, with Agamben, to the refugee as the "marginal figure" capable of unhinging "the old trinity of state-nation-territory." Agamben applies this question, like Said, to the question of Jerusalem:

> What is new in our time is that growing sections of humankind are no longer representable inside the nation-state – and this novelty threatens the very foundations of the latter....
>
>   One of the options taken into consideration for solving the problem of Jerusalem is that it become – simultaneously and without any territorial partition – the capital of two different states. The paradoxical condition of reciprocal extraterritoriality (or, better yet, aterritoriality) that would thus be implied could be generalized as a model of new international relations. ... Only in a world in which the spaces of states have been

thus perforated and topologically deformed and in which the citizen has been able to recognize the refugee that he or she is – only in such a world is the political survival of humankind today thinkable.[76]

"Only in such a world" – a world of such radically different worldliness that we would not recognize it as such. A world that solicits us to destroy the cultural-ecological correlation between national identity and territorialized space.

That is why the postcolonial critique of the worlding of the world requires us to finish the critique of cultural ecology. We should ask: what if there never was such a thing as a "culture area"? What would be left of cultural ecology, the scientific and ethnographic study of nature-culture linkages, the concepts of local and indigenous technical knowledges, and so forth? Do these questions have fundamental priority for our research? Or are they merely problems of method, to be dealt with by shifting our epistemological field or incorporating new variables, such as gender, class, or even history itself?

Critiques of cultural ecology in the 1980s spurred a transition from cultural to political ecology. Michael Watts' study of food production in northern Nigeria, *Silent violence*, which drew from cultural ecology and political economy to explain the politics of the reproduction of food economies, was fundamental for this shift.[77] Three departures from earlier work in cultural ecology, including Nietschmann's, are striking. The first concerns *scale*: whereas work like Nietschmann's assumed that cultural patterns were best examined at the scale of individual communities, Watts' intervention was to demand that cultural ecology be situated at multiple temporal and spatial scales simultaneously. This insight required the transformation of cultural ecology from a study of *systems* in isolation into a means of examining *practices* embedded in broader historical-geographical processes. The second change involves *politics*: Watts insisted that patterns of resource extraction and use could never be isolated as mere cultural-ecological phenomena, since they are intensely bound up with state policies, labor reproduction, and capitalist social relations. Third, whereas cultural ecology tended to treat the existence of systems *ahistorically*, as though the production of food in a particular ecological system was cultural by the fact of its inherent logic, Watts showed that this could not be the case by historicizing the colonial state's involvement in the often insecure systems of household food production in Nigeria. Notwithstanding these arguments, *Silent violence* did call for human or cultural ecology to be replaced as such.[78]

In hindsight, Watts' critique helps us to understand why cultural ecology inspired counter-mapping. Counter-mapping stands as the political demonstration of cultural ecology's essential truth. In light of our reading of the *Atlas*, we should recognize that Watts did not go far enough in his critique. It is time to deconstruct cultural ecology through the ceaseless reversal and spacing of the structured ideas of culture, nature, region, nation, and system at its core. Because the subaltern cannot map. Rather, she does: and yet when we are presented with subaltern maps we do not read them to disorient ourselves. We go on. Our course does not change. Or perhaps we do not know where else to go.

It is a measure of the hegemony of the colonial present to recognize that there is nothing inherently emancipatory in the desire for counter-maps. A counter-map that turns the tables on colonialism may produce a worlding that still turns within a colonial form of power. *And yet:* we must embrace the fact that we cannot *not* desire counter-mapping.[79] We should honor our postcolonial desire to overthrow the map of Europe. Therefore the task is as clear as it is aporetic: we must counter-map, and yet relentlessly critique those maps, always reading towards the concepts and strategies that will produce the strongest and most radically open, anticolonial modes of worlding the world. Our success will be measured by the degree to which the anticolonial texts that are produced do not fall into the sway of the power of territorialization by nation-states that use the counter-maps as a resource for capitalism qua development. It is too soon to know where the *Maya Atlas* will be swayed. Rereading the text in a postcolonial, deconstructive purview is a strategy for knowing where it leads us, for asking where we are to go.

Notes

1   In this chapter I do not follow the "counter-mapping" literature in speaking of "counter-hegemony." Gramsci did not use the expression "counter-hegemony." This language seems to have emerged in the late 1970s, but it is a misreading of Gramsci. "Counter" suggests "turning the tables" and is often equated with any act of anti-conformism. Gramsci specifically defines hegemonic struggle as struggle for moral and political leadership and/or state power. When subaltern groups successfully struggle in Gramscian terms, it is because they win hegemony, not because they act willy-nilly "counter hegemonic." One should substitute therefore "struggle for a different hegemony," but of course only when the activity qualifies under the Gramscian qualifications of the term. (I am indebted to J. Buttigieg on this point.)

2 A lot has transpired in this literature since 1995; see especially King (2002); Hodgson and Schroeder (2002); Herlihy and Knapp (2003); Bryan (2007); Kwan (2007).

3 Poole (1995: 1). The special issue is available at: www.culturalsurvival.org/publications/csq/index.cfm?id=18.4.

4 Nietschmann (1995: 37), italics mine.

5 The draft I worked on extended Assi's May 1995 "Proposal for mapping Maya land use in the Toledo District." It appears that Assi wrote this proposal with assistance from the Indian Law Resource Center. Although the concept of mapping Maya communities in Belize had been discussed by TMCC since at least 1985, the onset of the project in 1995 was a result of several contingent events. A USAID-funded conservation and development project, NARMAP, offered funding to the TMCC in 1995 to demarcate the boundaries of the Maya villages in Toledo. NARMAP's leaders felt that effective conservation in southern Belize would require clear and settled boundaries and saw the TMCC's project as a means to this end. But NARMAP's offer of financial support threatened to change the character of TMCC's initial designs, since NARMAP was interested in demarcating *village* boundaries, and not mapping and fighting for trans-community rights to land resources. The TMCC rejected NARMAP's offer and sought funds for a more ambitious project. These events coalesced around the efflorescence of the Maya movement in Belize, only two years after Rigoberta Menchú was awarded the Nobel Peace Prize, and a year after the Zapatista uprising in Chiapas. This is to remind us that the Maya mapping project was not simply imagined and initiated by people "within" Belize.

6 TMCC and TAA (1997: inside cover).

7 On the contingent articulation of national-popular ideology, see Hall (1996: 141–5).

8 TMCC and TAA (1997: 5).

9 Bol (1997).

10 See note 2. There have also been at least three well-attended conferences specifically on the theme, two special issues of academic journals, and a number of websites dedicated to indigenous counter-mapping.

11 Peluso (1995: 393); see also Hodgson and Schroeder (2002).

12 On "worldliness," see Heidegger (1996: Division one, Part III).

13 There is a close parallel here to a conceit, used by B. Q. Nietschmann in two of his books, that there are countries that we cannot see ("There is a country between Nicaragua and Cuba. It is not on any map..."). Elsewhere I have called the impossibility of properly naming colonized indigenous spaces "the aporia of postcolonial geography" (Lund and Wainwright, forthcoming).

14 The *Atlas* and counter-mapping stands in relation to conventional geography as subaltern or social history stands with respect to conventional historiography: it is a project aimed at restoring the full presence of the

other in language, or reclaiming the silenced voice of the subaltern in geography. Our task is to study the debates about subaltern history so that we do not repeat the pitfalls of that project. This raises the question of whether spatiality is somehow different from temporality in a way that changes what is at stake, or possible, in a postcolonial reading.

15   Spivak (1999: 306). See note 1.

16   Except for the final workshop, on layout. I also conducted background research; worked on logistical arrangements; facilitated communications between the TMCC and the alcaldes; collected GPS data for the base maps; helped to train the community researchers; edited text; and worked with the project leaders to compile and sort the quantitative data from the community survey.

17   Since 2003 I have been part of a group that has built a new indigenous-rights group, the Julian Cho Society (JCS), that aims to carry forward the work of the late Julian Cho. The JCS formally and publicly emerged in July, 2005, and now (September, 2007) employs a staff of three, based in an office in Punta Gorda – located in Julian Cho's old home. I am presently Vice-Chair of the JCS council and one of its more active members. We are working in about nine Maya communities (it depends how you number and count communities – another postcolonial aporia that troubles counter-mapping) to try to materialize the 2004 decision of the Inter-American Commission on Human Rights in favor of the Maya. This is the case that the *Atlas* was mapped for. Among other things, the JCS is presently working with Maya communities on a project mapping. So the desire for a counter-map has not been abated. Counter-maps have their own kind of hegemony. My position in JCS's internal debate has been to argue that border-mapping and community demarcation projects should not precede the forging of political strategies by local indigenous leaders. This chapter (originally written around the time of the formation of the JCS) explains some of my reasons for valuing mobilizing over geo-referencing.

18   Cited in Anon. (1996). Nietschmann made this statement in a speech at the *Atlas* cartography training sessions. I believe that I am the "anonymous" author of the press release that led to this story.

19   The map, entitled "Proposed Cultural Homeland of the Indigenous Members of Belize in Toledo," is reprinted on page 4 of the *Atlas*.

20   Cho was appointed Chair of the Committee by the alcaldes around August, 1995; I started working with Cho and the Committee in early October, 1995.

21   Cho was reelected in 1998. Cho's leadership between 1995 and his untimely death in November 1998 was critical to the success of these projects. By that time, the *Atlas* had been published; the logging in the CRFR had ended; the lawsuit had been filed with the Inter-American Commission on Human Rights (IACHR); and – after the Musa-led People's United Party took power in 1998 – the lawsuit had been set aside in favor

of friendly settlement talks with the state on the land question. Cho died under mysterious circumstances in November, 1998 and the friendly settlement talks soon collapsed, as the Maya movement subsided and the government failed to negotiate. In 2003 the IACHR decided in favor of the Maya. The struggle to transform this decision into concrete territorial gains for the Maya continues (see subsequent note).

22   The Supreme Court of Belize effectively ignored the case put forward by the Maya, and the government, in its role as defendant, limited its response to suggesting that the Maya were not indigenous to Belize at all, and were merely recent immigrants of Guatemala. In its 2004 ruling, the Inter-American Commission on Human Rights concluded otherwise, noting that the Maya could not only trace longstanding historical and cultural ties (thus qualifying as indigenous people under the norms of international indigenous rights law) but, moreover, that the state had committed harms against the Maya's human rights:

> The State violated the right to property enshrined in Article XXIII of the American Declaration to the detriment of the Maya people, by failing to take effective measures to recognize their communal property right to the lands that they have traditionally occupied and used, without detriment to other indigenous communities, and to delimit, demarcate and title or otherwise establish the legal mechanisms necessary to clarify and protect the territory on which their right exists.

> The State further violated the right to property enshrined in Article XXIII of the American Declaration to the detriment of the Maya people, by granting logging and oil concessions to third parties to utilize the property and resources that could fall within the lands which must be delimited, demarcated and titled or otherwise clarified and protected, in the absence of effective consultations with and the informed consent of the Maya people.

The IACHR concluded that the Government of Belize must:

> Adopt in its domestic law, and through fully informed consultations with the Maya people, the legislative, administrative, and any other measures necessary to delimit, demarcate and title or otherwise clarify and protect the territory in which the Maya people have a communal property right, in accordance with their customary land use practices, and without detriment to other indigenous communities.

23   Cho (1996).

24   As a publisher of alternative health, martial arts, and spiritual titles, the *Atlas* was a departure from North Atlantic's normal collection: www.northatlanticbooks.com.

25   I conducted research on the local distribution and effects of the text in 2001–2. I spoke with about thirty people who were involved in the project, including three formal interviews with project leaders. In the course of a survey of one Maya community, I also collected comments and criticisms

of the *Atlas*. I carried a copy of the *Atlas* and would share the text with the family at the end of the interview. This usually led to far-reaching discussions about the text, its histories, and the reasons that the *Atlas* was not available in the community.

26 This amounts to three days' wages for an unskilled laborer in southern Belize. In any event, copies of the *Atlas* are hard to find in Belize. According to the publisher, only between one and two thousand copies of the text were sold – mainly in the USA.

27 Coc (2002). On the origins of ESTAP, see Wainwright (1998: ch. 4).

28 While the *Atlas* played an important role in training and empowering a group of Maya men and distributed some capital into Maya communities in a fairly direct and equitable fashion, it also established a precedent that later caused friction within the Maya movement: because all of the people who worked on the mapping project were paid a *per dium* wage whether they worked a full day or not, the project established the standard that volunteers for TMCC and TAA would receive a cash payment for all labor. This has caused no small amount of difficulty for the often cash-poor Maya organizations.

29 TMCC and TAA (1997: 8–24).

30 Ibid., p. 18. A pasture is defined as "an area where cattle, horses, or mules are confined to graze." We should pause to ask: why are these terms defined? If neither "milpa" nor "pasture" are words from Maya languages, and if (in the case of pasture) its meaning is well-known, then what is the purpose of such a glossary? We begin to sense that the text aims to explain this system in a way that marks it as something that is both real and essential to the Maya, and also a thing that must be understood in a "technical" sense. This is suggested by the tension between the list of definitions of types of land use (which suggest an exact and translatable meaning for these complex cultural practices) and also the radically unspecified meanings of photos and stories concerning land use found elsewhere in the text.

31 The nature of these rules is not defined here, but this statement is clarified further in the sentence following the paragraph cited above: "The Maya do *milpa*, or slash-and-burn farming." *Milpa* is what the Maya do; it is the name for where *milpa* takes place; and it entails slashing and burning the forest.

32 We could then read this passage to mean that the government does not recognize the traditional rules for land management in Maya communities, which is true in the sense of legal recognition (giving consent to); but certainly the Maya farm system is recognized (see chapter 2).

33 The term "information" suggests mistranslation. The usual Q'eqchi' translation for "information" is *puktesil*. But *puktesil* is also translated for the word "news" to refer to "information" of a particular sort: it is information that tells one what has happened in a shallow, newsy way. It is not substantive knowledge.

34 Earlier in the *Atlas* we find a statement that suggests otherwise: "We are the original inhabitants of Toledo Belize who know no boundaries [*sic*]. The concept of putting down boundaries is European" (p. 2).

35 Ibid., p. 19.

36 Many Maya farmers in southern Belize no longer conduct such rituals.

37 Ibid., p. 21. For an ethnographic reading of Q'eqchi' corn-planting practices, see Wilson (1995: 110–12), where Wilson writes the following about the sexed rituals surrounding planting: "The atmosphere at the planting feast is especially heavy and pregnant. The men are shy and eat quietly, as if embarrassed or guilty... [since they] have just had 'sex' with the earth and fertilized it, and have now returned to their wives and the household."

38 TMCC and TAA (1997: 21).

39 Ibid., p. 20.

40 Ibid., p. 21. This photo, like the photo of the *milpa* on page 19, was not taken by one of the disposable cameras carried by the Maya photographers. Judging by their quality and style, I believe that both photos were taken by Nietschmann. I was present when he took the "corn storage" photo; he commented on the way the packed corn brought out interesting patterns and colors. This raises two points. First, during the making of the *Atlas*, it was often said that all of the photos in the book would be taken by the community researchers. This is not what actually happened, in part because the cheap, disposable cameras that were given to the researchers produced poor quality, grainy photos. The disposable cameras took only 36 photos, fewer than Nietschmann would take (on a professional-quality Nikon camera) every hour or so. Technology, access to film, training, and an "eye" for certain kinds of images thus contributed to a situation where Nietschmann's photos dominate the *Atlas*. Clearly this has implications for how we interpret the *Atlas* and Nietschmann's assertion that it is an essentially Maya work. Second, what makes the corn photo compelling is its composition, which suggests unity, order, organic structure, and abundance. There are no photos in the *Atlas* showing loose piles of corn, but it is not uncommon to find corn in piles in people's homes. So what makes Nietschmann's image of neatly arranged corn so "Maya"? How can *this* form be said to be "traditional," as opposed to simply "orderly"?

41 On indigeneity/modernity as a sign structure, see Gupta (1998).

42 This is consistent with all of the studies that have been conducted on the economic life of Maya households in southern Belize: see especially Wilk (1997).

43 Arguably the most significant strategy of resistance to state repression by the Q'eqchi' since the Spanish conquest has been to migrate – from their "cultural core" in Alta Verapaz to the Peten, Isabel, southern Belize, and elsewhere (Wilk 1997; Grandia 2006). Thus the abjection of the stories of migration constitutes a kind of forgetting of a political form of resistance.

44 Except for the cooking staff, all of the Maya people involved in the project were men. There were three white American women who worked on the project (two for only a short period and as assistants to Nietschmann). The woman with the most prominent role in the project, Deborah Schaaf, represented the Indian Law Resource Center in the project. Schaaf and I lobbied for greater inclusion of Maya women in the project.

45 During the making of the *Atlas* this section was referred to as "gender issues" or simply "women." The very existence of the "special topics" section is a clue that the "women's section" needed some home that the authors could not locate in the "normal" part of the text.

46 The making of jippi-jappa baskets probably dates to the 1980s. A 1972 study of pottery and basket-making in southern Belize found no such baskets produced in San Pedro Columbia or San Antonio (Hughes 1972).

47 See McClusky (2001), an ethnography of domestic violence among the Maya of southern Belize. While we should ask whether this ethnography actually naturalizes violence by giving it a cultural character, and whether it silences the women it claims to represent, the study nonetheless documents that spousal abuse is common practice used to humiliate and subordinate women. The silencing and objectification of women and girls in the *Atlas* could reinforce the belief that such violence is an essential character of Maya culture, but this would presume that men define "Maya culture."

48 Nietschmann's eye for such photographs was honed by work at *National Geographic*, which has a long tradition of producing documentary photos of women as sexual, traditional, native, natural objects. See Lutz and Collins (1993).

49 This tension rends the *Atlas* around the relationship between the documentary project of the *Atlas* and the desire that the documentation of Maya life will produce a "Maya Homeland." The *Atlas* reflects an ambivalent desire: on one hand, to *describe* the facts of existence for the Mayas and the land; and on the other, to articulate a vision of the Maya homeland. This is clear in its overarching goal – winning indigenous title to land – but also in the very grammar, mode of writing, and tone of the *Atlas*.

50 Heidegger (1996: section I.1.3).

51 Heidegger's attempt to analyze the metaphysics of presence in terms of the horizon of temporality only reiterates the Western metaphysical emphasis on time criticized by Derrida. I take it that this is one of the challenges Derrida proposes in his essay "Ousia and gramme", i.e., the task of spacing Heidegger. Derrida suggests in this essay that there may never have been such a thing as vulgar time – that this conceptual *différance* for Heidegger cannot be left undeconstructed. I am not so sure. If there is such a thing as vulgar time, it is manifested by the atomic clocks, manufactured and calibrated by US military scientists, that "keep time" for the network of GPS satellites that orbit the Earth today. These military satellites are the ultimate arbiters of real, exact, empirical space today, and they produce

their exact measurements through the precise measurement of *lag time*; hence the importance of their atomic clocks. Before the advent of GPS, Heidegger saw that the digitization of temporality was worlding the world in line with "the age of the world picture."

52 Heidegger (1996).

53 I return to this practice below. Whether or not this practice is unique to maps does not change my argument here, that the way to read a map is to focus on the relation between its *"point* of departure" and its particular representation of worldliness.

54 TMCC and TAA (1997: x).

55 The maps most commonly used for land use planning in Belize are from the 1:50,000 collection (the second map mentioned in the quote, above), which represents Belize on 44 sheets. The most recent aerial photos were taken in 1988 by the Royal Air Force at a scale of 1:48,000. The 1938, 1970, and 1988 aerial photos are archived by the Department of Lands and Surveys of the Ministry of Natural Resources. "Ground-truthing" was completed by the Belize Survey Department in 1974, with military grid numbering added in 1978. Before the 1:50,000 series, the standard maps used in British Honduras were the 1:39,304 Provisional series, drafted in 1952 for the colonial government. Of course, aerial photography has been largely replaced by space-based satellite imaging.

56 The category "aboriginal" and "indigenous" comes into existence through the arrival of colonial relations. In the colonial-indigenous encounter, it is always the former who name the latter as aborigines. The political reversal of these two terms in claims to justice does not change the concepts or their genealogies. On tracing the native informant, see Spivak (1999). Subjectivity is always spatial (not only when we speak of indigenous peoples), albeit in an indeterminate way: cf. Lacan's argument that language assigns the subject's "place" and Althusser's reading of interpellation. Regarding the latter: in the interpellative hailing by the police – "hey you, *there*" – we find the call of subjectivity taking place. Interpellation therefore always already works in a subject's sociospatial environment.

57 Nietschmann (1997: 1). Nietschmann's letter begins with best wishes to all "in the Maya Homeland" and mainly comprises a request for payment for services rendered to the TMCC and TAA.

58 Nietschmann (1973). Nietschmann conducted the research for his dissertation (which became *Between land and water*) at the University of Wisconsin-Madison in the late 1960s. His doctoral committee was chaired by Bill Denevan, a cultural ecologist and Berkeley geographer. Nietschmann's involvement in the political and intellectual ferment at Madison in the 1960s deeply informed his cultural ecology and intellectual trajectory. During this period Nietschmann came to identify with those nations that were struggling against US imperialism, especially including nations that were not recognized as such or which had not achieved statehood.

He also became interested in systems ecology, and he developed an approach to ecosystems, culture, and social life that could only be described as "systematizing."

59   Ibid., p. x.

60   Steward (1955: 31). The genealogy of cultural ecology is complex, as it brought together elements of Sauer's cultural-historical approach with notions of adaptation, evolution, and systems from postwar systems ecology (especially through Odum's ecological research).

61   Ibid., p. 37. For Steward, this analysis focuses on three elements: (1) "the interrelationship of exploitative or productive technology and environment"; (2) the "behavior patterns involved in the exploitation of a particular area by means of a particular technology"; and (3) the "extent to which the behavior patterns entailed in exploiting the environment affect other aspects of culture"; see Steward (1955: 40–1). Culture is here defined foremost through language and land use practices.

62   Ibid., p. 36.

63   Braun (2004).

64   This is not to say that the discourse of international indigenous rights law was not instrumental to the conceptualization and writing of the *Atlas*. It was: particularly in the logic to the assembling of the spatial data from the different communities into the two key maps drawn at a smaller scale. But it did not determine the form or narrative of the text.

65   Nietschmann (1994: 226).

66   Ibid., p. 226.

67   TMCC and TAA (1997: 2).

68   Zimmerer (2001).

69   Although Zimmerer is correct about the *Atlas*'s hybrid character, there is, as we have seen, much more to this story. The *Atlas* should not be read as another step towards ethnolandscape ecology – which would presuppose that the Mayas have/are a preexisting ethnos – to contribute to landscape reading – but rather as a text *of*, and calling *for*, postcolonial geography.

70   Nietschmann also asks for "Maya history in the region, the settlement history of the communities," i.e., the Mayas' historical memory as much as their transhistorical cultural artifacts. Yet his tone and the examples betray a sense in which "these things" that constitute "Maya life" are so self-evident that they need not be historicized.

71   Ibid., p. 2

72   This particular student had no qualms about inserting himself into the *Atlas*. During the final layout stage, he placed 13 photos of himself on the section about the making of the *Atlas* (only six pages long) – more than one finds of the project's three main leaders (Bol, Cho, and Nietschmann) combined.

73   Said (1996: 26).

74 Ibid., p. 27. Let me underscore that I completely agree with Said's critique of the PLO and Oslo. The problem lies precisely in that our desire for *our own national* maps and censuses cannot but reproduce the practices – at once theoretical, material, and discursive – and building blocks laid by European colonial states, including the map and the census: see Anderson (1983).

75 Said (1996: 31).

76 Agamben (2000: 20–1, 24, 26).

77 Watts (1983).

78 Watts writes at one point, "this is not the place to engage in a critique of human ecology" (ibid., p. 87), but the critique was implied throughout the 600 pages. Reflecting on *Silent violence* eighteen years after its publication, Watts writes:

> *Silent Violence* was an attempt to obviously move beyond the limitations of natural hazards research and cultural ecology: beyond functionalism, the pitfalls of behavioralism, and the reluctance to engage in the political economy of the market. In my own case, more than anything else this took the form of a head-on collision with the magnificent work of Roy Rappaport (my anthropology teacher at Michigan) and the cultural ecology of Barney Nietschmann [working down the hall from Watts at Berkeley]. Central to this engagement was an exploration of the ways in which the properties of living systems and their adaptive properties could be made to speak to the social relations of production understood in a Marxist sense. (Watts 2001b: 626)

79 I am presently involved in another indigenous counter-mapping project coordinated through the JCS (see note 17).

Conclusion

However, we are not concerned here with the condition of the colonies. The only thing that interests us is the secret discovered in the new world by the Political Economy of the old world, and proclaimed on the housetops: that the capitalist mode of production and accumulation, and therefore capitalist private property, have for their fundamental condition the annihilation of self-earned private property, in other words, the expropriation of the laborer.

Karl Marx (1992: 724)

Its execution directed at the question of capitalism as a mode of production, Marx's critique of political economy in the first volume of *Capital* had to unravel the question of the commodity. This is a condition for an analysis of value, the exploitation of labor by capital, and, at the very end of the book – *colonialism*.[1] Yet if we were to write an itinerary (since *Capital*) of the relationship between Marxisms defined vis-à-vis *value*, on one hand, and anti-imperial theories of *colonialism*, including postcolonialism, on the other, we would find no shortage of failed connections and miscommunications. These leave the end of *Capital* unhinged: there is a radical break, both in tone and substance, separating Marx's triumphal conclusion of chapter 32 – where the "entanglement of all peoples in the net of the world-market," rise of monopoly power, and "oppression," along with the "revolt of the working-class . . . at last reach a point where they become incompatible with their capitalist integument . . . [and] burst asunder," sounding "the knell of capitalist private property" – and the beginning of chapter 33, which is taken up with a cool critique of Wakefield's theory of colonialism. A Marxist-postcolonialist reading of the itinerary of the differential spaces of these two traditions must reconsider the relation between *property* and *territory*, between the social relations of property and the production of the state's space. In this brief chapter 33, Marx famously explains that "capital is not a thing, but a social relation between persons." Today the postcolonialist would insist the same of territory. Marxism has been largely defined from its origin by opposition to property as a social relation, whereas postcolonialism has emerged from the ruins of nationalist struggles against colonialism – what Said once called the great achievement of the twentieth century – which were predicated on the reconquest of national territory. Consolidating the theoretical gains of postcolonial Marxism, we should be able to conceive of colonial power as the simultaneous extension of territorial and capitalist social relations.[2]

Yet the different emphases placed on territory and property is not the only matter that has separated postcolonialism and Marxism. There are also the contentious problems of orientialism in the Marxist tradition and the complicity of imperialism sanctioned by actually existing socialist states. These should seem minor next to the points of collaboration and solidarity between Marxism and postcolonialism: their critiques of imperialism, of representation, and of nationalism. Yet, again, their *différance*, the matter of property/territory, is no small matter. How can we forge solidarity on these grounds?

This book has demonstrated that the answer centers, today, on the problem of *development*. It is development, after all, that both Marxism and anticolonialism have tried to provide to subaltern classes – and generally failed to deliver. And it is development that structures our conceptions of global space (developed-or-developing) and gives cover to the violence of capitalist social relations. Remember Chatterjee's conclusion: "the historical identity between Reason and capital has taken on the form of an epistemic privilege, namely, "*development*".[3] Today we cannot responsibly repeat Marx's gesture from the end of *Capital* and say that "we are not concerned here with the condition of the colonies." The postcolonial condition is one where nothing is of greater concern than "the condition of the colonies," that is, the living geography of the world's intense divisions between rich and poor. Because this condition is still addressed as a question *of* development, rethinking the historical identity between Reason and capital requires attending specifically to capitalism qua development. As I have shown through the analysis of the colonization and development of the Maya, we may take hold of development conceptually in such a way as to be able to think more critically of capitalism's great historical-geographical achievement – of *becoming* capitalism qua development. My purpose in writing this book was to show how this became possible in one place, but as a consequence of the planetary hegemony of capitalism qua development, no doubt this analysis could be repeated elsewhere.

But in the mean time, you may rightly ask, what exactly is to be done? Within the genre of development studies, nearly every book concludes by answering, explicitly or not, Lenin's famous question (though typically not in Lenin's register). Like Hollywood films that always end well, the tendency to cite illustrations of successful development (or worse, progress in development *studies*) reflects, I think, a facile dishonesty by suggesting that an easy way out of our immense difficulties lies right around the corner. This implies that development is, despite all the indications to the contrary, a simple matter to be resolved. It is not. There are no simple formulæ for "real" development, and a gesture toward a particular radical program as *the* solution to the question of development ("look at the Zapatistas") cannot do justice to the diverse contexts and substantive differences that make, for instance, the EZLN so remarkable.

If I refuse to offer any such directions here it is not for lack of respect for my reader or the importance of the task, nor for lack of ideas, but only to remain true to the mode of analysis I have used throughout this book. I am presently collaborating with Maya activists on radical, non-capitalist projects in southern Belize, but to cast these as alternatives to "development" would be dishonest, for they are not, and these current political struggles were not my object in this book. We should respect the distance between theory and political risk-taking even as we may try to overcome it.[4] A more honest and proximate gain would be for us to connect the dots between the conceptual destruction of capitalism qua development and what Gramsci called "the analysis of situations," i.e., a careful discerning of social class relations, organic-versus-conjunctural forces and tendencies, etc., that allow for strong political strategies. In this register it is difficult to say anything about what is happening generally with a responsible clarity: Gramsci, for his part, did not write in such terms; nor, for different reasons, did Derrida.

Owning up to such great challenges may be the most responsible way to conclude, but that is not my end here. Facing difficulties in writing about Beethoven's late style – certainly one of the most complex situations in all of Western music – even a writer as brilliant as Adorno was moved to admit:

> In the face of the late works...one has the feeling of something extra-ordinary, and of an extreme seriousness of a kind hardly to be found in

any other music. At the same time, one feels an uncommon difficulty is saying precisely – . . . in terms of the composition – in what this extra-ordinariness and seriousness actually consist. One therefore escapes into biography.[5]

But biography, of course, will not do, since it is precisely the *music*, not Beethoven, that must be analyzed. On this point I venture the following parallel: this is like saying it is capitalism qua development, not any particular illustration of alternatives, that matters most. The escape toward concrete examples of good strategies, or proper development, at the end of so many otherwise critical development studies reflects the escape to biography. And yet as Adorno writes of the *Missa Solemnis*, "the extraordinary difficulty which [this question] presents even to straightforward understanding should not deter us from interpret-ation."[6] The same must be said of the aporia of development.

I turn here to Adorno's magnificent essays on the late style of Beetho-ven's final compositions not only for this metaphor, but because Adorno offers us a wonderfully concrete way of rethinking the aporia of devel-opment, again in a passage on Beethoven's *Missa Solemnis*:

> But the question still remains why the late Beethoven, who must have stood very aloof from organized religion, devoted many years of his most mature period to a sacred work and – at the time of his most extreme subjective emancipation – experimented with the rigidly bound style. The answer seems to me to lie in direct line with Beethoven's critique of the "classical' symphonic ideal. The bound style *allows* him a development which is hardly permitted by instrumental music.[7]

Adorno proposes seven reasons that explain why Beethoven's style in the *Missa* "*allows* him" this unpermitted and unanticipated development. The first of these is: "there are no tangible themes – and therefore *no development*."[8]

What can this mean? An unanticipated development derives from the absence of development? A great, unpermitted, unanticipated develop-ment – from no development?

Adorno's insight suggests a way of thinking through the aporia of development, a way of thinking development that begins where it ends. Adorno indicates that one of the conditions that allows Beethoven's radical, unprecedented development (i.e., the sort of development that we may want) lies in its total lack of *thematic* development. This allows us to ask: what are the tangible themes of development? What themes must be detached from development to permit an unanticipated, novel development? What would we need to destroy to allow us, in Derrida's words, to think and put development "to work otherwise"?[9]

As this book has shown, the dominant, tangible themes of development today are capitalism, settling, and trusteeship. These themes, inherited from Western colonialism, dominate development. To think anew, to think toward unanticipated forms of development, requires releasing development from these themes. To carry this out will require an experimental postcolonial Marxism. The unity of capitalism and development in its twofold sense, capitalism qua development, was forged in colonialism, and sustains imperial power. We have yet to unsettle this power and to break capitalism's attachment to development. Such is the challenge of decolonizing development.

Notes

1   Marx (1992: 716–24). I have imitated the form of Derrida's opening lines from "Ousia and gramme: Note on a note from *Being and time*" (1982: 31). (See chapter 6, note 51.)

2   Ismail (2006).

3   Chatterjee (2001: 169), my italics.

4   This breach is represented by the semicolon of Marx's famous thesis on Feuerbach: "The philosophers have only *interpreted* the world, in various ways; the point is to *change* it" (1992: 423). The latter clause does not cancel the first. Marx underscores the need for both interpretation and change, as well as the distinction between them, even as he celebrates change.

5   Adorno (1998: 138).

6   Ibid., italics in original. Beethoven's *Missa Solemnis* is Opus 123.

7   Ibid., pp. 138–9.

8   Ibid., p. 139. Elsewhere Adorno writes that Beethoven's late style "still remains process, but not as development" (cited in Said 2006: 10).

9   Derrida (1994: 90).

Bibliography

Archival Sources

Unpublished archival sources are cited in the footnotes using the following abbreviations:

Public Record Office (PRO), now the National Archives, Kew, England.

Archives of Belize (AB), Belmopan, Belize. *Minute Papers* are noted "AB, MP"; *Miscellaneous Collection* texts "AB, MC."

The archives of the TRDP and TSFDP project are labeled "Tumul K'in," as they were in the care of *Tumul K'in* until they were burned.

Works Cited

Adorno, T. 1998 [1962]. "Progress." In *Critical models: Interventions and catchwords*. New York: Columbia University Press.

Adorno, T. 1998. *Beethoven: The philosophy of music*. Stanford, CA: Stanford University Press.

Agamben, G. 1998. *Homo sacer: Sovereign power and bare life*. Stanford, CA: Stanford University Press.

Agamben, G. 2000. *Means without end: Notes on politics*. Trans. V. Binetti and C. Casarino. Minneapolis: University of Minnesota Press.

Agamben, G. 2005 [1990]. *The coming community*. Minneapolis: University of Minnesota Press.

Agnew, J. and S. Corbridge. 1995. *Mastering space: Hegemony, territory and international political economy*. New York: Routledge.

Aguirre, R. 2005. *Informal empire: Mexico and Central America in Victorian culture*. Minneapolis: University of Minnesota Press.

Ahmad, A. 1992. *In theory: Classes, nations, literatures*. New York: Verso.

Aldana, E. and P. Lee. 1982. "Rice production data." Unpublished manuscript. Punta Gorda, Belize: TRDP.

Alder, D. 1993. *Report of an assessment of broadleaf forest resource and sustainable yield in Belize*. Belmopan, Belize: Forest Planning and Management Project.

Alexander, D. 2001. *Mayanism*. Pinellas Park, FL: Sentient Temple.

Althusser, L. 1971. "Ideology and ideological state apparatuses (notes towards an investigation)." In *Lenin and philosophy and other essays*. Trans. B. Brewster. New York: Monthly Review Press.

Amin, S. 1974. *Unequal development*. New York: Monthly Review Press.

Amin, S. 1989 [1988]. *Eurocentrism*. Trans. R. Moore. New York: Monthly Review Press.

Amin, S. 1990. *Maldevelopment*. London: Zed Books.

Anderson, B. 1983. *Imagined communities*. New York: Verso.

Anon. 1921. *Colony of British Honduras*. London: Waterlow & Sons.

Anon. 1934. "Economic products from British Honduras." *Bulletin of the Imperial Institute* 32: 356–375.

Anon. 1937. *Report of the Forest Department British Honduras for the year 1936*. Belize: Forest Department of British Honduras.

Anon. 1946. *Empire forests and the war*. London: Miscellaneous Official Publications No. 62.

Anon. 1996. "Maya cartographers commence mapping." *Eco* 1(2): 38.

Anon. 2000. *George Price: Father of the nation Belize*. Belize City, Belize: ION Media.

Arias, A. ed. 2001. *The Rigoberta Menchú controversy*. Minneapolis: University of Minnesota Press.

Aristotle. 1941. *Physics*. In *The basic works of Aristotle*, Trans. & ed. R. McKeon. New York: Modern Library.

Aristotle. 1979. *The politics*. Trans. & ed. T. A. Sinclair. London: Penguin.

Arnason, J. T. and J. D. H. Lambert. 1982. "Nitrogen cycling in the seasonally dry forest zone of Belize, Central America." *Plant and Soil* 67: 333–342.

Ashcraft, N. 1973. *Colonialism and underdevelopment: Processes of political economic change in British Honduras*. New York: Columbia University Press.

Aspinall, A. E. 1924. *British Honduras*. London: Dunstable & Watford and Waterlow & Sons.

Aspinall, A. E. 1926. *The handbook of the British West Indies, British Guiana and British Honduras, 1926–27*. London: West India Committee.

Asturias, M. A. 1973. "Forword." In P. Ivanoff, *Maya*. New York: Madison Square Press.

Asturias, M. A. 1977 [1923]. *Guatemalan sociology: The social problem of the Indian*. Trans. M. Ahern. Tempe: Arizona State University Press.

Asturias, M. A. 1993 [1949]. *Men of Maize [Hombres de maíz]*. Trans. G. Martin. Pittsburgh, PA: University of Pittsburgh Press.

Avery, W. 1900. "British Honduras." *British American Geographical Society*: 331–333.

Barillas, E. 1988. *El "problema del indio" durante la epoca liberal.* Guatemala: Universidad de San Carlos.

Barnes, T. 1996. *Logics of dislocation: Models, metaphors, and meanings in economic space.* New York: Guilford Press.

Barthes, R. 1986. *The rustle of language.* Trans. R. Howard. New York: Hill & Wang.

Bartlett, H. 1935. *A method of procedure for field work in tropical American phytogeography based upon a botanical reconnaissance in parts of British Honduras and the Peten forest of Guatemala.* Washington, DC: Carnegie Institute.

Baskerville, C. 1984. *Matahambre corn: The production of maize during the off-season.* Punta Gorda, Belize: TRDP.

Baskerville, C. 1984. *Production of beans in the uplands of Southern Toledo.* Punta Gorda, Belize: TRDP.

Bataillon, M. 1951. "La Vera Paz, roman et histoire." *Bulletin Hispanique* 53: 235–300.

Bataillon, M. 1971. "The Clérigo Casas, colonist and colonial reformer." In *Bartolomé de Las Casas in history: toward an understanding of the man and his work.* Ed. J. Friede and B. Keen. DeKalb: University of Northern Illinois Press: 353–440.

Belize Information Service. 1996. Untitled press release. Belmopan, Belize: January 23, 1996.

Bellamy, J. 1889. "Expedition to the Cockscomb Mountains, British Honduras." *Proceedings of the Royal Geographical Society and Monthly Record of Geography,* 11(9): 542–552.

Berkey, C. 1994. "Maya land rights in Belize and the history of Indian reservations." Washington, DC: Indian Law Resource Center.

Berte, N. 1983. *Agricultural production and labor investment strategies in a K'ekchi' village: Southern Belize.* PhD Diss., Department of Anthropology. Chicago, IL: Northwestern University.

Bhabha, H. 1990. *Nation and narration.* New York: Routledge.

Bhabha, H. 1994 [1992]. "The other question: stereotype, discrimination, and the discourse of colonialism." In *The location of culture.* New York: Routledge, 66–85.

Biermann, B. 1971. "Bartolomé de Las Casas and Verapaz." In *Bartolomé de Las Casas in history: Toward an understanding of the man and his work.* Ed. J. Friede and B. Keen. DeKalb: University of Northern Illinois Press, 443–484.

Bird, N. 1994. "Draft Forest Management Plan: Columbia Forest Management Unit." Belmopan, Belize: FPMP.

Bird, N. 1995. "Regarding 'Mayan Homeland In Belizean Homeland Under Siege By Malaysian Loggers.'" Unpublished manuscript. Belmopan, Belize: FPMP.

Bird, N. 1998. *Sustaining the yield: Improved timber harvesting practices in Belize*. Chatham, UK: Natural Resources Institute.

Black, J. F. 1998. "Scorched earth in a time of peace." *NACLA* 32(1): 11–15.

Blaikie, P. 2000. "Development, post-, anti-, and populist: A critical review." *Environment and Planning A* 32(6): 1033–1050.

Bol, D. 1997. "Why the Maya Atlas?" Unpublished manuscript (speech). Belmopan, Belize.

Bolland, O. N. 1977. *The formation of a colonial society: Belize, from conquest to Crown Colony*. Baltimore, MD: Johns Hopkins University Press.

Bolland, O. N. 1979. "The Maya and the colonization of Belize in the nineteenth century." In *Anthropology and history in Yucatan*. Ed. G. Jones. Austin: University of Texas Press.

Bolland, O. N. 1986. *Colonialism and resistance in Belize: Essays in historical sociology*. Benque Viejo, Belize: Cubola.

Bolland, O. N. 1987. "Alcaldes and Reservations: British policy towards the Maya in late nineteenth century Belize." *America Indigena* 47: 33–76.

Bolland, O. N. 1997. *Struggles for freedom: Essays on slavery, colonialism, and culture in the Caribbean and Central America*. Belize City, Belize: Angelus.

Bolland, O. N. and A. Shoman 1977. *Land in Belize 1765–1871*. Mona, Jamaica: Institute of Social and Economic Research, University of the West Indies.

Bourbon, F. 2000. *The lost cities of the Mayas: The life, art, and discoveries of Frederick Catherwood*. New York: Abbeville.

Braun, B. 2000. "Producing vertical territory: Geology and governmentality in late Victorian Canada." *Ecumene* 7: 7–46.

Braun, B. 2002. *The intemperate rainforest: Nature, culture, and power on Canada's west coast*. Minneapolis: University of Minnesota Press.

Braun, B. 2004. "Nature and culture: On the career of a false problem." In *A companion to cultural geography*. Ed. J. Duncan, N. Johnson, and R. Schein. Oxford: Blackwell, 151–179.

Braun, B. and J. Wainwright. 2001. "Nature, poststructuralism, and politics." In *Social nature: Theory, practice, and politics*. Ed. N. Castree and B. Braun. Oxford: Blackwell, 41–63.

Brenner, N. 1999. "Beyond state-centrism? Space, territory, and geographical scale in globalization studies." *Theory and Society* 28(1): 39–78.

Breton, A. 1988. "Los indios Mayas de Belize." *Trace* (13): 32–34.

British Museum. 1938. *Guide to the Maudslay collection of Maya sculptures (casts and originals) from Central America*. Oxford: Oxford University Press.

Brotherston, G. 1996. "Review of Prieto, *Miguel Angel Asturias's Archeology of Return*." *MLN* 111(2): 430–432.

Bryan, J. 2007. *Map or be mapped: Land, race, and rights in eastern Nicaragua*. PhD Diss., Department of Geography. University of California-Berkeley.

Burdon, J. 1927. *Brief sketch of British Honduras*. London: West India Committee.

Burdon, J. 1935. "Historical note." In *Archives of British Honduras*, Vol. 1. Ed. J. Burdon. London: Sifton Praed.

Burdon, J. ed. 1935. *Archives of British Honduras*, Vols. 1–3. London: Sifton Praed.

Butler, J. 1993. *Bodies that matter*. New York: Routledge.

Butler, J. 2000. "Dynamic conclusions." In J. Butler, E. Laclau, and S. Žižek. *Contingency, hegemony, universality: contemporary dialogues on the left*. London: Verso.

Caiger, S. L. 1935. *Westward Ho! A glimpse at the diocese of British Honduras, Central America*. London: SPG & SPCK.

Caiger, S. L. 1951. *British Honduras past and present*. London: George Allen & Unwin.

Cambranes, J. C. 1985. *Coffee and peasants: The origins of the modern plantation economy in Guatemala*. Woodstock, VT: CIRMA.

Camille, M. 1994. "Government initiative and resource exploitation in Belize." PhD Diss. Department of Geography. Texas A&M University.

Carmichael, E. 1973. *The British and the Maya*. London: British Museum.

Carmody, P. 2001. *Tearing the social fabric: Neoliberalism, deindustrialization, and the crisis of governance in Zimbabwe*. Portsmouth, NH: Heinemann.

Carozza, P. 2003. "From conquest to constitutions: Reviving a Latin American tradition of human rights." *Human Rights Quarterly* 25: 281–313.

Carr, A. 1922. "British Honduras." *British Pan American Union* 54: 262–274.

Casteñeda, Q. 1996. *In the museum of Maya culture: Touring Chichén Itzá*. Minneapolis: University of Minnesota Press.

Castree, N. 2005. *Nature*. New York: Routledge.

Castree, N. and B. Braun. eds. 2001. *Social nature: Theory, practice, and politics*. Oxford: Blackwell.

Caufield, C. 1996. *Masters of illusion: The World Bank and the poverty of nations*. New York: Henry Holt.

Chakrabarty, D. 2000. Provincializing Europe: Postcolonial thought and historical difference. Princeton, NJ: Princeton University Press.

Chapin, M. and B. Threlkeld. 2001. *Indigenous landscapes: A study of ethnocartography*. Arlington, VA: Center for the Support of Native Lands.

Chatterjee, P. 2001 [1986]. *Nationalist thought and the colonial world: A derivative discourse?* Minneapolis: University of Minnesota Press.

Chaturvedi, V. ed. 2000. *Mapping subaltern studies and the postcolonial*. New York: Verso.

Chibber, V. 2006. *Locked in place: State-building and late industrialization in India*. Princeton, NJ: Princeton University Press.

Cho, J. 1995. Untitled press release from the Toledo Maya Cultural Council. Punta Gorda, Belize.

Cho, J. 1996. Statement on Channel 5 News, April 11, 1996. Belize City: Channel 5.

Cho, J. 1999. "Julian Cho honored by Audubon Society." Belize City: Channel 5.

Clark, G. 1936. *The balance sheets of imperialism: Facts and figures on colonies*. New York: Columbia University Press.

Clegern, W. 1967. *British Honduras: Colonial dead end*. Baton Rouge: Louisiana State University Press.

Coc, P. 1998. Letter to the Editor. *Belize Times*. Belize City.

Coc, P. 2002. Personal communication (interview).

Coc, P. and R. Cayetano. 1997. "An open letter to Esquivel." Punta Gorda, Belize.

Cockburn, S. 1875. *Rough notes and official reports on the River Belize, the physical features of British Honduras, taken in 1867 & 1869*. Kingston, Jamaica: C. L. Campbell.

Coe, M. 1977. *The Maya*. New York: Praeger.

Collet, W. 1909. *British Honduras and its resources*. London: West India Committee.

Conklin, H. 1962. "An ethnoecological approach to shifting agriculture." In *Readings in cultural geography*. Ed. P. Wagner and M. Mikesell. Chicago, IL: University of Chicago Press.

Constantine, S. 1984. *The making of British colonial development policy, 1914–1940*. London: Frank Cass.

Cook, J. 1935. *Remarks on a passage from the River Balise, in the Bay of Honduras, to Merida: The capital of the Province of Jucatan in the Spanish West Indies*. Providence, RI: John Carter Brown University Library, 1–34.

Cook, O. F. 1919. "Milpa agriculture, a primitive tropical system." *Annual report of the Smithsonian Institution for 1919*. Washington, DC: Smithsonian Institute, 307–26.

Corbridge, S. 1998. "Beneath the pavement only soil: The poverty of post-development." *Journal of Development Studies* 34(6): 138–148.

Cowen, M. and R. Shenton. 1996. *Doctrines of development*. London: Routledge.

Cowgill, U. 1962. "An agricultural study of the Southern Maya lowlands." *American Anthropologist* 64(1): 273–286.

Crowe, F. 1850. *The gospel in Central America*. London: Charles Gilpin.

Crush, J. ed. 1995. *Power of development*. New York: Routledge.

Darcel, F. 1954. "A history of agriculture in the colony of British Honduras." Unpublished manuscript. AB, MC 2660.

Davis, M. 2006. *Planet of slums*. New York: Verso.

Derrida, J. 1972. *Positions*. Chicago, IL: University of Chicago Press.

Derrida, J. 1978 [1966]. "Sign, structure, and play in the discourse of the human sciences." In *Writing and difference*. Trans. A. Bass. Chicago, IL: University of Chicago Press.

Derrida, J. 1978. *Writing and difference*. Chicago, IL: University of Chicago Press.

Derrida, J. 1982. *Of spirit*. Chicago, IL: University of Chicago Press.

Derrida, J. 1982. "Ousia and gramme: Note on a note from *Being and time*." In *Margins of philosophy*. Chicago, IL: University of Chicago Press, 29–67.

Derrida, J. 1992. "Force of law: The mystical foundation of authority." In *Deconstruction and the possibility of justice*. New York: Routledge, 3–67.

Derrida, J. 1994. *Spectres of Marx: The state of the debt, the work of mourning, and the New International*. New York: Routledge.

Derrida, J. 1995. *The gift of death*. Trans. D. Wills. Chicago, IL: University of Chicago Press.

Diamond, J. 2003. "The last Americans: Environmental collapse and the end of civilization." *Harper's Magazine* 306(1837): 43–51.

Diamond, J. 2005. *Collapse: How societies choose to fail or succeed*. New York: Viking.

Environment, Social, Technical Assistance Project (ESTAP). 2000. *Regional Development Plan for Southern Belize*. Punta Gorda, Belize: ESTAP.

de Escobar, Padre A. 1841. "Account of the province of Vera Paz, in Guatemala, and of the Indian settlements or Pueblos established therein." *Journal of the Geographical Society* 11: 89–97.

Escobar, A. 1994. *Encountering development: The making and the unmaking of the Third World*. Princeton, NJ: Princeton University Press.

Esquivel, M. 1995. "Prime Minister's address to the nation." Belmopan, Belize: Government Printer.

Evans, G. 1948. *Report of the British Guiana and British Honduras settlement commission*. London: Her Majesty's Stationary Office.

Evans, P. 1995. *Embedded autonomy: States and industrial transformation*. Princeton, NJ: Princeton University Press.

Fanon, F. 1998 [1963]. *The wretched of the earth*. Oxford: Blackwell.

Ferguson, J. 1990. *The anti-politics machine: "Development", depoliticization and bureaucratic state power in Lesotho*. Minneapolis: University of Minnesota Press.

Foucault, M. 1972. *The archaeology of knowledge*. New York: Random House.

Foucault, M. 1973. *The birth of the clinic: An archaeology of medical perception*. New York: Vintage.

Foucault, M. 1979 [1977]. *Discipline and punish: The birth of the prison*. New York: Vintage.

Foucault, M. 1983. "How to lead an anti-fascist life: Preface to G. Deleuze and F. Guattari, *Anti-Oedipus*." In *Power: Essential works of Foucault*, Vol. 3. Ed. J. Faubion. New York: New Press, 106–110.

Foucault, M. 1994. *The order of things: An archaeology of the human sciences*. New York: Vintage.

Foucault, M. 1996. *Foucault live: Collected interviews, 1961–84*. Ed. S. Lotringer. New York: Semiotext(e).

Foucault, M. 1998 [1967]. "Different spaces." In *Michel Foucault: Aesthetics, method, and epistemology*, Vol. 2. Ed. J. Faubion. New York: New Press, 175–185.

Foucault, M. 2003 [1976]. *Society Must Be Defended*. New York: Picador.

Forest Department. 1995. "The facts about the Columbia River Reserve logging concession." Unpublished manuscript. Belmopan, Belize: November 7, 1995.

Forest Department. 1996. "GOB response to Belize Audubon Society's (BAS) release of 22nd May, 1996 on logging concession." Belmopan, Belize.

Forest Planning and Management Program [FPMP]. 1997. "Belize timber licenses, production and royalties." Belmopan, Belize: FPMP.

Fowler, H. 1879. *A narrative of a journey across the unexplored portion of British Honduras, with a short sketch of the history and resources of the colony.* Belize City: Government Press.

Friede, J. and B. Keen. eds. 1971. *Bartolomé de Las Casas in history: Toward an understanding of the man and his work.* DeKalb: University of Northern Illinois Press.

Gabb, A. 1992. *The impact of financial liberalization in Belize: 1974–1989.* PhD Diss. New York: New School for Social Research.

Gage, T. 1958. *Thomas Gage's travels in the New World.* Norman: University of Oklahoma Press.

Galeano, E. 1967. *Guatemala: Occupied country.* New York: Monthly Review Press.

Galeano, E. 2000. *Upside down.* New York: Picador.

Galeano, E. 2001. "Let's shoot Rigoberta." In *The Rigoberta Menchú controversy.* Ed. A. Arias. Minneapolis: University of Minnesota Press, 99–102.

Gallego y Cadena, V. 1962. "Precursores de los estudios ethnograficos en Guatemala: Vera Paz." *Guatemala Indigena* 2(3): 141–161.

Gallenkamp, C. 1976. *Maya: The riddle and rediscovery of a lost civilization.* New York: David McKay.

Gann, T. 1924. *In an unknown land.* London: Duckworth.

Gann, T. 1925. *Mystery cities: Exploration and adventure in Lubaantún.* New York: Scribner's Sons.

Gann, T. 1926. *Ancient cities and modern tribes.* New York: Scribner's Sons.

Gann, T. and J. E. S. Thompson. 1937. *The history of the Maya from the earliest times to the present day.* New York: Scribner's Son.

Garcia, J. 1999. "Report of the forest license review committee." Unpublished manuscript. Belmopan, Belize.

Gibbs, A. R. 1883. *British Honduras: An historical and descriptive account of the Colony from its settlement, 1670.* London: Sampson Low, Marston, Searle, & Rivington.

Gidwani, V. 2002. "The unbearable modernity of 'development'? Canal irrigation and development planning in Western India." *Progress in Planning* 58: 1–80.

Gidwani, V. Forthcoming. *Capital, interrupted: Development, agrarian change and the politics of work in Central Gujarat, India.* Minneapolis: University of Minnesota Press.

Glaser, M. 1995. "Why most Maya people in Toledo do not support the Columbia Forest Management Plan." Unpublished manuscript. Belmopan, Belize: FPMP.

Glassman, J. 2004. *Thailand at the margins: Internationalization of the state and the transformation of labour.* New York: Oxford University Press.

Glassman, J. and A. I. Samatar. 1997. "Development geography and the Third-World state." *Progress in Human Geography* 21(2): 164–198.

Gould, G. 1984 [1976]. "Of Mozart and related matters: Glenn Gould in conversation with Bruno Monsaingeon." In *The Glenn Gould reader.* Ed. T. Page. New York: Vintage.

Gramsci, A. 1957. *The modern prince and other writings.* New York: International Publishing.

Gramsci, A. 1971. *Selections from the prison notebooks.* New York: International Publishing.

Gramsci, A. 1992. *Antonio Gramsci: Prison notebooks,* Vol. 1. Ed. J. A. Buttigieg. New York: Columbia University Press.

Gramsci, A. 1995. *The Southern question.* West LaFayette, IN: Bordighera.

Gramsci, A. 1996. *Antonio Gramsci: Prison notebooks,* Vol. 2. Ed. J. A. Buttigieg. New York: Columbia University Press.

Grandia, L. 2006. *Unsettling: Land dispossession and enduring inequity for the Q'eqchi' Maya in the Guatemalan and Belizean frontier colonization process.* PhD Diss., Department of Anthropology. University of California-Berkeley.

Grandin, G. 2000. *The blood of Guatemala.* Raleigh, NC: Duke University Press.

Grandin, G. 2004. *The last colonial massacre: Latin America in the Cold War.* Chicago, IL: University of Chicago Press.

Grant, C. 1976. *The making of modern Belize: Politics, society and British colonialism in Central America.* New York: Cambridge University Press.

Gravin, R. 1973. *The crystal skull: The story of the mystery, myth and magic of Hitchell-Hedges crystal skull discovered in a Lost Mayan City during a search for Atlantis.* New York: Doubleday.

Green, E. 1997. Personal communication (interview). Belmopan, Belize.

Greene, A. 2003. "The voice of Ix Chel: Fashioning Maya tradition in the Belizean rain forest." In *In search of the rainforest.* Ed. C. Slater. Durham, NC: Duke University Press, 101–130.

Greene, A. 2005. *Extracting Tradition: Nature, Culture, and Power in the Ethnobotony of Belize.* PhD Diss., Department of Anthropology, University of California–Davis.

Gugliotta, G. 2007. "The Maya: Glory and Ruin." *National Geographic* 212(2): 68–109.

Gupta, A. 1998. *Postcolonial developments: Agriculture in the making of modern India.* Durham, NC: Duke University Press.

Hall, J. 1973. "Mennonite agriculture in a tropical environment; an analysis of the development and productivity of a mid-latitude agricultural system in British Honduras." PhD dissertation, Department of Geography. Worcester, MA: Clark University.

Hall, S. 1996 [1986]. "On postmodernism and articulation: An interview with Stuart Hall." *Journal of Communication Inquiry* 10(2): 145–160.

Hammond, N. 1972. *Lubaantún: 1926–1970*. London: Trustees of the British Museum.

Hammond, N. 1978. "The myth of the milpa: Agricultural expansion in the Maya lowlands." In *Pre-Hispanic Maya agriculture*. Ed. P. D. Harrison and B. L. Turner. Albuquerque: University of New Mexico Press, 23–34.

Handy, J. 1984. *Gift of the devil: A history of Guatemala*. Boston, MA: South End Press.

Hanke, L. 1953. "Bartolomé de Las Casas and the Spanish empire in America: Four centuries of misunderstanding." *Proceedings of the American Philosophical Society* 97(1): 26–30.

Hanke, L. 1959. *Aristotle and the American Indians: A study in race prejudice in the modern world*. Bloomington: Indiana University Press.

Hardt, M. and A. Negri. 2000. *Empire*. Cambridge, MA: Harvard University Press.

Harrison, P. 1978. "So the seeds shall grow: Some introductory comments." In *Pre-Hispanic Maya agriculture*. Ed. P. D. Harrison and B. L. Turner. Albuquerque: University of New Mexico Press, 1–11.

Harrison, P. and B. L. Turner. eds. 1978. *Pre-Hispanic Maya agriculture*. Albuquerque: University of New Mexico Press.

Harshbeerger, J. W. 1893. "Maize; a botanical and economic study." *University of Pennsylvania: Contributions from the Botanical Laboratory* 1(2).

Harss, L. and B. Dohmann. 1967. "Miguel Ángel Asturias, or the land where the flowers bloom." In *Into the mainstream: Conversations with Latin American authors*. New York: Harper & Row.

Hart, G. 2001. "Development critiques in the 1990s: Cul de sac and promising paths." *Progress in Human Geography* 25(4): 649–658.

Harvey, D. 1996. *Justice, nature, and the geography of difference*. Oxford: Blackwell.

Harvey, D. 2003. *The new imperialism*. Oxford: Oxford University Press.

Havinden, M. and D. Meridith. 1993. *Colonialism and development: Britain and its tropical colonies, 1850–1960*. London: Routledge.

Hegel, G. 1985. "Introduction." In *Lectures on the History of Philosophy*. Oxford: Oxford University Press.

Heidegger, M. 1972. *On Time and Being*. New York: Harper & Row.

Heidegger, M. 1977a. "The question concerning technology." In *The question concerning technology and other essays*. New York: Harper & Row, 3–35.

Heidegger, M. 1977b. "The age of the world picture." In *The question concerning technology and other essays*. New York: Harper & Row, 115–154.

Heidegger, M. 1996 [1927]. *Being and time*. Trans. J. Stambaugh. Albany: State University of New York Press.

Henderson, G. 1811. *An account of the British settlement of Honduras; being a view of its commercial and agricultural resources, soil, climate, natural history, and etc.* London: Baldwin.

Henighan, S. 1999. *Assuming the light: The Parisian literary apprenticeship of Miguel Ángel Asturias.* Oxford: Legenda.

Herlihy, P. and G. Knapp. 2003. "Maps of, by, and for the peoples of Latin America." *Human Organization* 62(4): 303–314.

Hervik, P. 1999. *Mayan people within and beyond boundaries: Social categories and lived identity in Yucatán.* Amsterdam: Harwood.

Hester, J. 1953. "Maya agriculture." *Carnegie Institution of Washington Yearbook* 53: 297–298.

Hitchens, C. 1998. *The Elgin marbles: Should they be returned to Greece?* New York: Verso.

Hobsbawm, E. 1989. *The age of empire: 1875–1914.* New York: Vintage.

Hodgson, D. L. and R. A. Schroeder. 2002. "Dilemmas of counter-mapping community resources in Tanzania." *Development and Change* 33: 79–100.

Howard, L. and C. Koehl. 1997. "More logging?" *Newsweek* (January 20) 129(3): 4.

Howard, M. 1974. "Agricultural labor among the Indians of the Toledo District." *National Studies* 2: 1–13.

Howard, M. 1992. *A Hegel dictionary.* Oxford: Blackwell.

Hughes, H. 1972. "Pottery and related handicrafts of San Antonio and San Pedro Colombia, Toledo District, Belize, British Honduras." Cambridge: University of Cambridge Centre of Latin American Studies Working Papers.

Hummel, C. 1921. *Report on the forests of British Honduras, with suggestions for far reaching forest policy.* London: British Colonial Research Committee.

Interdepartmental Committee on Maya Welfare [ICMW]. 1941. *Report of the Interdepartmental Committee on Maya Welfare.* Belize City: Government of British Honduras. AB, MC 1819.

Ismail, Q. 2006. *Abiding by Sri Lanka: On peace, place, and postcoloniality.* Minneapolis: University of Minnesota Press.

Ismail, Q. 2007. "Is culture a 'construct'? On the emergence of a colonial concept." Unpublished manuscript.

de Janvry, A. 1981. *The agrarian question and reformism in Latin America.* Baltimore, MD: Johns Hopkins University Press.

Jessop, B. 1982a. "Accumulation, state, and hegemonic projects." *Kapitalistate* 10/11: 89–111.

Jessop, B. 1982b. *The capitalist state: Marxist theories and methods.* Oxford: Blackwell.

Jessop, B. 1990. *State theory: Putting capitalist states in their place.* University Park: Pennsylvania State University Press.

Johnston, R. 1995. "Territory." In *Dictionary of human geography.* Ed. R. Johnston, D. Gregory, and D. Smith. Oxford: Blackwell.

Jones, G. 1982. "Agriculture and trade in the colonial period Southern Maya lowlands. Maya subsistence." In *Studies in memory of Dennis Puleston*. Ed. K. Flannery. New York: Academic Press, 275–293.

Jones, G. 1998. *The conquest of the last Maya kingdom*. Stanford, CA: Stanford University Press.

Joyce, T. A., T. Gann, et al. 1927. "Report on the British Museum Expedition to British Honduras, 1927." *Journal of the Royal Anthropological Institute of Great Britain and Ireland* 58: 295–333.

Karatani, K. 2000. "Uses of aesthetics: After orientalism." In *Edward Said and the work of the critic*. Ed. P. Bové. Durham, NC: Duke University Press, 139–151.

Karatani, K. 2005 [2001]. *Transcritique: On Kant and Marx*. Boston, MA: MIT Press.

Keck, M. and K. Sikkink. 1998. *Activists beyond borders*. New York: Cornell University Press.

Keen, B. ed. 1967. *Readings in Latin American civilization*. DeKalb, IL: Houghton Mifflin.

Keen, B. 1971. "Introduction: Approaches to Las Casas, 1535–1970." In *Bartolomé de Las Casas in history: Toward an understanding of the man and his work*. Ed. J. Friede and B. Keen. DeKalb, IL: University of Northern Illinois Press, 3–63.

Keen, B. 1977. "The legacy of Bartolomé de las Casas." *Ibero-Americana Pragensia* (Prague) 11(5): 7–67.

Kempton, J. H. 1931. "Maize, the plant breeding achievement of the American Indian." *Smithsonian Scientific Series* 11: 1–3.

King, B. 2002. "Towards a participatory GIS: Evaluating case studies of participatory rural appraisal and GIS in the developing world." *Cartography and Geographic Information Science* 29(1): 43–52.

Kothari, U. 2005. "From colonial administration to development studies: A postcolonial critique of the history of development studies." In *A radical history of development studies*. Ed. U. Kothari. New York: Zed Books, 47–66.

Kroshus Medina, L. 1999. "History, culture, and place-making: 'Native' status and Maya identity in Belize." *Journal of Latin American Anthropology* 4(1): 133–165.

Kwan, M. 2007. "Affecting geospatial technologies: Toward a feminist politics of emotion." *Professional Geographer* 59(1): 22–34.

Lalu, P. 2005. "The grammar of domination and the subjection of agency: Colonial texts and modes of evidence." *History and Theory* 39: 45–68.

de Las Casas, B. 1954 [1537]. *Real cédula afirmando la contrata que hizo el gobernador de la provincia de Guatemala con Fray Bartolomé de Las Casas, sobre ir con otros religiosos a la pacificacíon de ciertos indios reldes*. Trans. L. Hanke and M. Giménez Fernández. Santiago de Chile.

de Las Casas, B. 1954 [1539]. *Real cédula a Alvarado, gobernador de Guatemala*. Trans. L. Hanke and M. Giménez Fernández. Santiago de Chile, 56.

de Las Casas, B. 1958 [1540]. *Carta al Emperador. Opusculos, Cartas y Memoriales*. Madrid, 68–69.

de Las Casas, B. 1967 [1909]. *Apologética historia sumaria de las Indias*, Vol. 1. Mexico: UNAM.

de Las Casas, B. 1992 [ca. 1537]. *The only way: A translation of Del único modo de atraer a todos los pueblos a la verdadera religión*. Trans. F. P. Sullivan. Ed. H. R. Parish. New York: Paulist Press.

de Las Casas, B. 1992 [1542]. *A short account of the destruction of the Indies*. Trans. N. Griffin. New York: Penguin.

Latouche, S. 1993. *In the wake of the affluent society: An exploration of post-development*. London: Zed Books.

Latour, B. 1987. *Science in action: How to follow scientists and engineers through society*. Cambridge, MA: Harvard University Press.

Laws, G. 1928. "The survey of the Lubaantun district in British Honduras: Paper read at the Meeting of the Society on 5 December 1927 by Major Cooper Clark in the absence of the author." *Geographical Journal* 71: 224–239.

Lawson, K. 1992. "An inexhaustible abundance: The national landscape depicted in American magazines, 1780–1820." *Journal of the Early Republic* 12(3): 303–330.

Lee Peluso, N. 1995. "Whose woods are these? Counter-mapping forest territories in Kalimantan, Indonesia." *Antipode* 27(4): 383–406.

Lefebvre, H. 2002 [1978]. "Space and the state." In *State/space: A reader*. Ed. N. Brenner, B. Jessop, M. Jones, and G. Macleod. Oxford: Blackwell, 84–100.

Lehmann, D. 1997. "An opportunity lost: Escobar's deconstruction of development." *Journal of Development Studies* 33(4): 568–578.

Lemmon, A. 1973. "San Luis of San Antonio." *National Studies* 1(6): 21–23.

Lenin, V. I. 1997 [1916]. *Imperialism: The highest stage of capitalism, a popular outline*. New York: International Publishing.

Li, T. 1999. "Compromising power: Development, culture and rule in Indonesia." *Cultural Anthropology* 14(3): 295–322.

Lovell, W. G. 1988. "Surviving conquest: The Maya of Guatemala in historical perspective." *Latin American Research Review* 23(2): 25–60.

Lovell, W. G. 1990. "Mayans, missionaries, evidence, and truth: The polemics of native resettlement in sixteenth-century Guatemala." *Journal of Historical Geography* 16(3): 277–294.

Lucas, C. 1905. *A historical geography of the British Colonies*. Oxford: Clarendon Press.

Lund, J. and J. Wainwright. Forthcoming. "Miguel Ángel Asturias and the aporia of postcolonial geography." *Interventions: A Journal of Postcolonial Theory*.

Lundell, C. L. 1933. "The agriculture of the Maya." *Southwest Review* 19: 65–77.

Lundell, C. L. 1945. "The vegetation and natural resources of British Honduras in plants and plant science in Latin America." In *Chronica Botanica*. Ed. F. Verdoorn. Waltham, MA, 270–273.

Lutz, C. and J. Collins. 1993. *Reading National Geographic*. Chicago, IL: University of Chicago Press.

McCaffrey, C. 1967. Potentialities for community development in a Kekchi Indian village in British Honduras. PhD Dissertation in Education, University of California.

McCalla, W. 1995. "Draft paper on possibilities for local revenue generation arising from commercial timber extraction under the Columbia Forest Reserve Management Plan." Unpublished report to the Belize Forest Department. Belmopan, Belize.

McClintock, A. 1992. "The angel of progress: Pitfalls of the term 'post-colonialism.'" *Social Text* 31/32: 84–98.

McClusky, L. 2001. *Here, life is hard: Stories of domestic violence from a Mayan community in Belize*. Austin: University of Texas Press.

McCreery, D. 1994. *Rural Guatemala, 1760–1940*. Stanford, CA: Stanford University Press.

Marrouchi, M. 2000. "Counternarratives, recoveries, refusals." In *Edward Said and the work of the critic: Speaking truth to power*. Durham, NC: Duke University Press, 187–228.

Martin, G. 1993. "Introduction." In *Men of maize*. Trans. G. Martin. Pittsburgh, PA: University of Pittsburgh Press, xi–xxxii.

Marx, K. 1906 [1887]. *Capital*, Vol. 1. Trans. S. Moore and E. Gaveling. New York: Modern Library.

Marx, K. 1978. "The Eighteenth Brumaire of Louis Bonaparte." In *The Marx-Engels reader*. Ed. R. Tucker. New York: Norton, 594–617.

Marx, K. 1992 [1887]. *Capital*, Vol. 1. New York: International Publishing.

Mason, J. A. 1927. "What we know about the Maya." *Museum Journal* 18(4): 351–380.

Melville, T. and M. Melville. 1971. *Guatemala: The politics of land ownership*. New York: Free Press.

Mitchell, T. 1988. *Colonizing Egypt*. Berkeley, CA: University of California Press.

Mitchell, T. 1991. "America's Egypt: Discourse of the development industry." *Middle East Report* 21(2): 18–36.

Mitchell, T. 1992. "Going beyond the state? A response to critiques." *American Political Science Review* 86.

Mitchell, T. 1998. "Fixing the economy." *Cultural Studies* 12(1): 82–101.

Mitchell, T. ed. 2000. *Questions of modernity*. Minneapolis: University of Minnesota Press.

Mitchell, T. 2003. *Rule of experts: Egypt, techno-politics, modernity*. Berkeley: University of California Press.

Moberg, M. 1992. "Continuity under colonial rule: The alcalde system and the Garifuna in Belize." *Ethnohistory* 19(1): 1–19.

Moberg, M. 1996. "Crown Colony as banana republic: The United Fruit Company in Banana Honduras, 1900–1920." *Journal of Latin American Studies* 28(2): 357–381.

Molina, L. 1884. "Informe dirijido [*sic*] al Ministerio de Gobernación por el Jefe Político y Comandante de Armas del Departamento de la Alta Verapaz, Guatemala, General de División Don Luis Molina, acerca de su administración política y militar durante el año de 1884." Guatemala: Tipografía de P. Arenales.

Montejo, V. 2005. *Maya intellectural renaissance: Identity, representation, and leadership*. Austin: University of Texas Press.

Moore, D. 1999. "The crucible of cultural politics: Reworking 'development' in Zimbabwe's eastern highlands." *American Ethnologist* 26(3): 654–689.

Moore, D. 2005. *Suffering for territory: Race, place, and power in Zimbabwe*. Durham, NC: Duke University Press.

Moore, R. E. 1967. *Historical dictionary of Guatemala*. Metuchen, NJ: Scarecrow Press.

Moreiras, A. 2000. "Ten notes on accumulation." *Interventions* 2(3): 343–363.

Moreno-Lopez, P. 1997. *Inter-American Development Bank Report: Belize*. Washington, DC: Inter-American Development Bank.

Morley, D. and K. Chen. eds. 1996. *Stuart Hall: Critical dialogues in cultural studies*. New York: Routledge.

Morley, S. G. 1946. *The ancient Maya*. Stanford, CA: Stanford University Press.

Morris, D. 1883. *The Colony of British Honduras, its resources and prospects, etc.* London: E. Stanford.

Mowitt, J. 2005. *Re-takes: Postcoloniality and foreign film languages*. Minneapolis: University of Minnesota Press.

Musa, S. 2003. "No turning back: The budget speech for fiscal year 2003/ 2004." Belmopan, Belize: Government of Belize.

Namis, N. and M. Glaser. 1992. "Toledo Uplands fieldwork findings." Unpublished manuscript. Belmopan, Belize: FPMP.

Nash, J. 2001. *Mayan visions*. New York, Routledge.

Naylor, R. A. 1967. "Guatemala: Indian attitudes toward land tenure." *Journal of Inter-American Studies* 9(4): 619–639.

Nelson, D. 2000. *A finger in the wound*. Berkeley: University of California Press.

Nietschmann, B. 1973. *Between land and water: The subsistence ecology of the Miskito Indians, eastern Nicaragua*. New York: Seminar.

Nietschmann, B. 1979. *Caribbean edge: The coming of modern times to isolated people and wildlife*. New York: Bobbs-Merrill.

Nietschmann, B. 1979. "Ecological change, inflation, and migration in the far western Caribbean." *Geographical Review* 69(1): 1–24.

Nietschmann, B. 1989. *The unknown war: The Miskito nation, Nicaragua, and the United States*. New York: Freedom House & University.

Nietschmann, B. 1993. "Nicaragua's new environmental alliance for Indian-Latin America: Research and exploration." *National Geographic* 9: 270–271.

Nietschmann, B. 1994. "The Fourth World: Nations versus states." In *Reordering the world: Geopolitical perspectives on the twenty-first century.* Ed. G. Demko and W. Wood. New York: Seminar, 225–242.

Nietschmann, B. 1995. "Defending the Miskito reefs with maps and GPS." *Cultural Survival Quarterly* 18(4): 34–37.

Nietschmann, B. 1997. Letter to J. Cho, L. Acal, and D. Schaaf, February 19. Unpublished manuscript.

Nuñez, F. 1977. "Administration and culture: Subsistence and modernization in Crique Sarco, Belize." *Caribbean Quarterly* 23(4): 17–46.

Olson, J. D. 1991. *The Indians of Central and South America: An ethnohistorical dictionary.* Westport, CT: Greenwood Press.

Osborn, A. 1982. Socio-anthropological aspects of development in Southern Belize. Unpublished manuscript. Punta Gorda, Belize: TRDP.

Overseas Development Administration [ODA]. 1989. "Belize Tropical Forestry Action Plan." London: Overseas Development Administration.

Owen-Lewis, D. 1998. "A Kekchi odyssey." Unpublished manuscript.

Owen-Lewis, D. 2001. Personal communication (interview), July 1.

Pacheco, L. 1985. *Religiosidad Maya-Kekchi' alrededor del maíz.* San Jose, Costa Rica: Editorial Escuela para Todos.

Palacio, J. 1996. "Development in Belize, 1960–1980." *Belizean Studies* 22(2/3).

Parker, T., B. K. Holst, et al. 1993. *A biological assessment of the Columbia River Forest Reserve, Toledo District.* Belize: Conservation International.

Penados, F. 2006. "Tumul K'in Center of Learning Strategic Plan 2006–2011." Unpublished manuscript. Blue Creek, Belize: Tumul K'in.

Pickles. J. 1993. "Texts, hermeneutics and propaganda maps." In *Writing worlds: Discourse, text, and metaphor in the representation of landscape.* Ed. T. Barnes and J. Duncan. New York: Routledge.

Pieterse, J. 2000. "After post-development." *Third World Quarterly* 21(2): 175–191.

Pim, A. W. 1933. *British Honduras: Financial and economic position.* London: His Majesty's Stationary Office.

Poole, P. 1995. "Land-based communities, geomatics and biodiversity conservation." *Cultural Survival Quarterly* 18.4 (January 31): 1–4.

Poulantzas, N. 2000 [1979]. *State, power, socialism.* Trans. P. Camiller. London: New Left Books.

Pratt, M. L. 2001. "*I, Rigoberta Menchú* and the 'culture wars.'" In *The Rigoberta Menchú controversy.* Ed. A. Arias. Minneapolis: University of Minnesota Press, 29–48.

Radcliffe, S. 2005. "Development and geography II: Towards a postcolonial development geography?" *Progress in Human Geography* 29(3): 291–298.

Rambo, A. T. 1962. "The Kekchi Indians of British Honduras: An ethnographic study." *Katunob Occasional Publications in Mesoamerican Anthropology* (5): 40–48.

Redfield, R. 1934. *Chan Kom.* Washington, DC: Carnegie Institute.

Reina, R. 1967. "Milpas and milperos: Implications for prehistoric times." *American Anthropologist* 69: 1–20.

Reina, R. and R. M. Hill. 1980. "Lowland Maya subsistence: Notes from ethnohistory and ethnography." *American Antiquity* 45(1): 74–79.

De Remesal, A. 2002. *Bartolomé de Las Casas (1474–1566) in the pages of Father Antonio de Remesal.* Trans. F. Jay. Lewiston, NY: Edwin Mellen Press.

Restall, M. 2003. "A reevaluation of the authenticity of Fray Diego de Landa's *Relación de las cosas de Yucatán.*" *Ethnohistory* 49(3): 651–669.

Rist, G. 1997. *The history of development.* New York: Zed Books.

Rodriguez, C. J. 1967. "Conquista de Verapaz." *Misionalia Hispanica* 70: 53–116.

Romney, D. H. 1962. "The fertility of lowland soils of British Honduras." *Empire Journal of Experimental Agriculture* 30: 95–107.

Roy, A. 2004. "Instant-mix imperial democracy." In *An ordinary person's guide to empire.* Boston, MA: South End Press, 41–68.

Said, E. 1979. *Orientalism.* New York: Vintage.

Said, E. 1981. *Covering Islam: How the media and the experts determine how we see the rest of the world.* New York: Pantheon.

Said, E. 1993. *Culture and imperialism.* New York: Vintage.

Said, E. 1994. "1994 afterword." In *Orientalism.* New York: Pantheon, 329–352.

Said, E. 1996 [1993]. "Facts, facts, and more facts." In *Peace and its discontents: Essays on Palestine in the Middle East peace process.* New York: Vintage.

Said, E. 2002 [1980]. "Tourism among the dogs." In *Reflections on exile.* Cambridge, MA: Harvard University Press, 93–97.

Said, E. 2002 [1988]. "Representing the colonized: Anthropology's interlocutors." In *Reflections on exile.* Cambridge, MA: Harvard University Press, 293–316.

Said, E. 2002 [1991]. "The politics of knowledge." In *Reflections on exile.* Cambridge, MA: Harvard University Press, 372–385.

Said, E. 2002 [1993]. "Culture and imperialism: An interview with J. Buttigieg and P. Bové." In *Power, politics, and culture.* New York: Vintage.

Said, E. 2002 [1995]. "History, literature, and geography." In *Reflections on exile.* Cambridge, MA: Harvard University Press, 453–473.

Said, E. 2003. "The imperial bluster of Tom DeLay: Dreams and delusions." *Counterpunch* (www.counterpunch.com) August 20.

Said, E. 2006. *On late style: Music and literature against the grain.* New York: Pantheon.

Samatar, A. I. 1989. *The state and rural transformation in Northern Somalia, 1884–1986*. Madison: University of Wisconsin Press.

Samatar, A. I. 1999. *An African miracle: State and class leadership and colonial legacy in Botswana development*. Portsmouth, NH: Heinemann.

Sampson, H. C. 1929. Report on development of agriculture in British Honduras. London: Her Majesty's Stationary Office.

Sanders, R. 2000. "Bernard Q. Nietschmann, 1941–2000." Unpublished manuscript. Downloaded February 5: www.geography.berkeley.edu/peoplehistory/Nietschmann/NietschmannMemorial.html.

Sapper, K. ed. 2000. "Early scholars' visits to Central America." *UCLA Occasional Papers*. Los Angeles: UCLA.

Scott, J. 1987. *Weapons of the weak: Everyday forms of peasant resistance*. New Haven, CT: Yale University Press.

Sharpe, J. and G. C. Spivak. 2002. "A conversation with Gayatri Chakravorty Spivak: Politics and the imagination." *Signs: Journal of Women in Culture and Society* 28: 609–624.

Shohat, E. 1992. "Notes on the 'post-colonial.'" *Social Text* 31/32: 99–113.

Shoman, A. 1987. *Party politics in Belize*. Belize City, Belize: Cubola.

Shoman, A. 1988. "Belize: An authoritarian democratic state in Central America." *SPEAR Reports* (4): 42–63.

Shoman, A. 1994. *Thirteen chapters of a history of Belize*. Belize City: Angeles Press.

Shoman, A. 1995 [1988]. *Backtalking Belize*. Belize City, Belize: Angeles Press.

Shoman, A. 1995. "Belize: An authoritarian democratic state." In *Backtalking Belize: Selected Writings*. Ed. A. Macpherson. Belize City: Angeles Press, 189–219.

Sidaway, J. 2000. "Postcolonial geographies: An exploratory essay." *Progress in Human Geography* 24: 591–612.

Simpson, J. A. and Weiner, S. C. eds. *The Oxford English dictionary*, 2nd edn. Oxford: Oxford University Press.

Smith, N. 1991. *Uneven development: Nature, capital, and the production of space*. Oxford: Blackwell.

Smith, N. 2003. *American empire*. Berkeley: University of California Press.

Solares, J. ed. 1990. *Estado y nacion: Las demandas de los grupos etnicos en Guatemala*. Guatemala: FLACSO.

Spivak, G. Chakravorty 1988. "Can the subaltern speak?" In *Marxism and the interpretation of culture*. Ed. Cary Nelson and Lawrence Grossberg. Urbana: University of Illinois Press.

Spivak, G. Chakravorty 1993. "An interview with Gayatri Chakravorty Spivak." *Boundary 2* 20:2: 24–50.

Spivak, G. Chakravorty 1994. "Responsibility." *Boundary 2* 21(3): 19–64.

Spivak, G. Chakravorty 1996. *The Spivak reader*. Ed. D. Landry and G. MacLean. New York: Routledge.

Spivak, G. Chakravorty 1999. *A critique of postcolonial reason: Toward a history of the vanishing present.* Cambridge, MA: Harvard University Press.

Spivak, G. Chakravorty 2006. "Thinking about Edward Said: Pages from a memoir." *Critical Inquiry* 31: 519–525.

Spivak, G. Chakravorty and J. Sharpe. 2002. "A conversation with Gayatri Spivak: Politics and the imagination." *Signs* 28(2): 609–624.

Squire, E. 1870. *Honduras: Descriptive, historical, and statistical.* London.

Standley, P. and S. Record. 1936. *The forests and flora of British Honduras.* Chicago, IL: Field Museum of Natural History.

Stephens, J. 1963 [1843]. *Incidents of travel in Yucatan,* Vols. 1 and 2. New York: Dover.

Steward, J. 1955. "The concept and method of cultural ecology." In *Theory of cultural change: The methodology of multilinear evolution.* Urbana: University of Illinois Press, 30–43.

Stockdale, F. 1932. "Agriculture in British Honduras." London: Colonial Advisory Council Paper No. 128.

Stoddart, D. 1965. "Geography and the ecological approach." *Geography* 50: 242–251.

Stoll, D. 1999. *Rigoberta Menchú and the story of all poor Guatemalans.* Boulder, CO: Westview Press.

Stone, M. 1995. "La Politica Cultural de la Identidad Maya en Belice." *Mesoamerica* 29: 162–214.

Strong, W. D. 1947. "Review of 'The Ancient Maya' by S. G. Morley." *American Anthropologist* 49: 640–641.

Swett, C. 1868. *A trip to British Honduras and to San Pedro, Republic of Honduras.* New Orleans, LA: George Ellis.

Thompson, J. E. S. 1938. "Sixteenth and Seventeenth Century Reports on the Chol Mayas." *American Anthropologist* 40: 585–564.

Thompson, J. E. S. 1966. *The rise and fall of Maya civilization.* Oklahoma City: University of Oklahoma Press.

Thompson, J. E. S. 1988 [1970]. *The Maya of Belize: Historical chapters since Columbus.* Belmopan, Belize: Benex Press.

Thompson, J. E. S. 1992. *The Maya of Belize: Historical chapters since Columbus.* Belmopan, Belize: Benex Press.

Toledo Maya Cultural Council [TMCC] and Toledo Alcaldes Association [TAA]. 1995. "Conditions to sustainably log a fraction of the Columbia River Forest Reserve and Indian lands." Unpublished manuscript. Punta Gorda, Belize.

Toledo Maya Cultural Council [TMCC] and Toledo Alcaldes Association [TAA]. 1996. "The San Jose Declaration." Unpublished manuscript. San Jose, Belize.

Toledo Maya Cultural Council [TMCC] and Toledo Alcaldes Association [TAA]. 1997. *Maya Atlas: The struggle to preserve Maya land in southern Belize.* Berkeley, CA: North Atlantic.

Turner, B. L. 1978. "The development and demise of the swidden thesis of Maya agriculture." In *Pre-hispanic Maya agriculture*. Ed. P. D. Harrison and B. L. Turner. Albuquerque: University of New Mexico Press, 13–22.

Turner, B. L. and P. D. Harrison. 1978. "Implications from agriculture for Maya prehistory." In *Pre-hispanic Maya agriculture*. Ed. P. D. Harrison and B. L. Turner. Albuquerque: University of New Mexico Press, 337–373.

Van Ausdal, S. 2001. "Development and discourse among the Maya of Southern Belize." *Development and Change* 32(3): 15.

Vandergeest, P. and N. Lee Peluso. 2001. "Territorialization and state power in Thailand." In *The legal geographies reader*. Ed. N. Blomley, D. Delaney, and R. Ford. Oxford: Blackwell, 177–186.

Viana, F. and F. L. Gallego, et al. 1962 [1574]. "Relación sobre la Provincia y tierra de Vera Paz (1544–1574)." *Guatemala Indigena* 2(3): 141–160.

von Hagen, V. W. 1973. *Search for the Maya: The story of Stephens and Catherwood*. Farnborough, UK: Saxon House.

Wade, R. 2004. "Is globalization reducing poverty and inequality?" *World Development* 32(4): 567–589.

Wainwright, J. 1998. *The political ecology of land in southern Belize*. MA Thesis, Department of Geography. University of Minnesota Press.

Wainwright, J. 2005. "The geographies of political ecology: After Edward Said." *Environment and planning A* 37(6): 1033–1043.

Wainwright, J. 2007. "First Affidavit of Joel David Wainwright." Unpublished manuscript. Presented to the Supreme Court of Belize, March 2007.

Wainwright, J. and C. Ageton. 2005. "The Cockscomb in the colonial present." *Journal of Belizean Studies* 27(2): 27–42.

Wainwright, J. and M. Robertson. 2003. "Territorialization, science, and the postcolonial state: The case of Highway 55 in Minneapolis, USA." *Cultural Geographies* 10: 197–217.

Wainwright, J., S. Prudham, and J. Glassman. 2000. "The battles in Seattle: Microgeographies of resistance and the challenge of building alternative futures." *Environment and planning D: Society and space* 18(1): 5–13.

Warren, K. 1998. *Indigenous movements and their critics: Pan-Maya activism in Guatemala*. Princeton, NJ: Princeton University Press.

Warren, K. 2001. "Pan-Mayanism and the Guatemalan peace process." In *Globalization on the ground: Postbellum Guatemalan democracy and development*. Ed. C. Chase-Dunn, S. Jonas, and N. Amaro. Lanham, MD: Rowman & Littlefield, 145–166.

Watts, M. 1983. *Silent violence: Food, famine, and peasantry in northern Nigeria*. Berkeley: University of California Press.

Watts, M. 1993. "Development I: Power, knowledge, discursive practice." *Progress in Human Geography* 17(2): 257–272.

Watts, M. 1999. "Collective wish images: Geographical imaginaries and post-development." In *Human Geography Today*. Ed. D. Massey and J. Allen. Cambridge: Polity Press, 85–107.

Watts, M. 2001a. "Development ethnographies." *Ethnography* 2(2): 283–300.

Watts, M. 2001b. "Author's response: Lost in space." *Progress in Human Geography* 25(4): 625–628.

Widdifield, S. 1990. "Dispossession, assimilation, and the image of the Indian in late-nineteenth-century Maxican painting." *Art Journal* 49(2): 125–132.

Wilk, R. 1981. "Agriculture, ecology, and domestic organization among the Kekchi Maya." PhD Diss. Department of Anthropology. University of Arizona.

Wilk, R. 1985. "Dry season riverbank agriculture among the Kekchi Maya, and its implications for prehistory." In *Prehistoric Lowland Maya Environment and Subsistence Economy*. Ed. Mary Pohl. Papers of the Peabody Museum Vol. 77, 47–58.

Wilk, R. 1986. "Mayan ethnicity in Belize: A historical review." *Cultural Survival Quarterly* 10(2): 73–77.

Wilk, R. 1987. "The Kekchi and the settlement of Toledo District." *Belizean Studies* 15(3): 33–49.

Wilk, R. 1987. "The search for tradition in Southern Belize: A personal narrative." *America Indigena* 47(1): 77–93.

Wilk, R. 1997. *Household ecology: Economic change and domestic life among the Kekchi Maya in Belize*. DeKalb: Northern Illinois University Press.

Wilk, R. 2002. "Free to be you and me?" Unpublished paper, presented at the conference "Towards an Ethical Mayan Archaeology," University of British Columbia, November, 2002.

Wilk, R. 2006. *Home cooking in the global village: Caribbean food from buccaneers to ecotourists*. New York: Berg.

Wilken, G. 1971. "Food-producing systems available to the ancient Maya." *American Antiquity* 36: 432–448.

Willard, T. 1933. *The lost empires of the Itzaes and Mayas: An American civilization, contemporary with Christ, which rivaled the culture of Egypt*. Glendale, CA: Arthur Clark.

Willey, G. R. 1978. "Pre-hispanic Maya agriculture: A contemporary summation." In *Pre-hispanic Maya agriculture*. Ed. P. D. Harrison and B. L. Turner. Albuqurque: University of New Mexico Press, 325–335.

Williams, R. 1983. *Keywords: A vocabulary of culture and society*. New York: Oxford University Press.

Wilson, R. 1995. *Maya resurgence in Guatemala: Q'eqchi' experiences*. Norman: University of Oklahoma Press.

Wolfgang, H. V. 1960. *World of the Maya*. New York: New American Library.

Wolfgang H. V. 1973. *Search for the Maya: The story of Stephens and Catherwood*. Farnborough, UK: Saxonhouse.

World Bank. 1979. *Current economic position and prospects of Belize*. Washington, DC: World Bank.

World Bank. 2003. *World development report 2003*. Washington, DC: World Bank.

Wright, A. C. S. ca. 1960. "Queen Elizabeth's milpa." Unpublished manuscript. AB, BAD/CHW 19.

Wright, A. C. S. ca. 1994. "The digging stick (thought about enhanced integration of the Toledo District with the rest of Belize that will follow from up-grading of the Southern Highway." Unpublished manuscript. AB, BAD/CHW/31, 2 pp.

Wright, A. C. S. 1995. "A history of the Columbia River Forest Reserve." Unpublished manuscript. AB, MC 5229.

Wright, A. C. S. 1996a. "The fifty million dollar question." Unpublished manuscript. AB, BAD/CHW/223.

Wright, A. C. S. 1996b. "Projects intended to modify Mayan land use in Toledo, 1953–1996." Unpublished manuscript. AB, BAD/CHW/82.

Wright, A. C. S. 1997a. "The use of land by Maya Indians." Unpublished manuscript. AB, BAD/CHW 15.

Wright, A. C. S. 1997b. Personal communication (interview), June.

Wright, A. C. S., D. Romney, R. Arbuckle, and V. Vial. 1959. *Land in British Honduras: Report of the Land Use Survey Team*. London: Her Majesty's Stationary Office.

Ximenez, F. 1965. *Historia de la Provincia de San Vicente de Chiapa y Guatemala de la Orden de Prdicadores*. Guatemala: CIRMA.

Yáñez, A. 2001 [1942]. *Fray Bartolomé de Las Casas: El conquistador conquistado*. México: D. F. Planeta.

Zimmerer, K. 2001. "Geographical review essay: Report on geography and the new ethnobiology." *Geographical Review* 91(4): 725–734.

Index

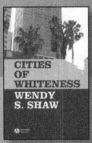

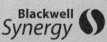